SOCIOPHOBIA

INSURRECTIONS

CRITICAL STUDIES IN RELIGION,
POLITICS, AND CULTURE

INSURRECTIONS: CRITICAL STUDIES
IN RELIGION, POLITICS, AND CULTURE

SLAVOJ ŽIŽEK, CLAYTON CROCKETT,
CRESTON DAVIS, JEFFREY W. ROBBINS, EDITORS

The intersection of religion, politics, and culture is one of the most discussed areas in theory today. It also has the deepest and most wide-ranging impact on the world. Insurrections: Critical Studies in Religion, Politics, and Culture will bring the tools of philosophy and critical theory to the political implications of the religious turn. The series will address a range of religious traditions and political viewpoints in the United States, Europe, and other parts of the world. Without advocating any specific religious or theological stance, the series aims nonetheless to be faithful to the radical emancipatory potential of religion.

For a complete list of books in this series, see page 169

CÉSAR RENDUELES

TRANSLATED BY
HEATHER CLEARY

SOCIOPHOBIA

POLITICAL CHANGE IN
THE DIGITAL UTOPIA

COLUMBIA UNIVERSITY PRESS

NEW YORK

Columbia University Press
Publishers Since 1893
New York   Chichester, West Sussex
cup.columbia.edu

Sociofobia was first published in Spanish in 2013 by Capitan Swing,
Rafael Finat 58, 28044 Madrid (SPAIN).
Rights negotiated by Oh! Books Literary Agency (info@ohbooks.es)
© 2013 César Rendueles
Translation copyright © 2017 Columbia University Press
All rights reserved

Library of Congress Cataloging-in-Publication Data
Names: Rendueles, César, author.
Title: Sociophobia : political change in the digital utopia / César
Rendueles; translated by Heather Cleary.
Other titles: Sociofobia. English
Description: New York : Columbia University Press, [2017] |
Series: Insurrections: critical studies in religion, politics, and culture |
Includes bibliographical references and index.
Identifiers: LCCN 2016041870 (print) | LCCN 2017000836 (ebook) |
ISBN 9780231175265 (cloth : alk. paper) | ISBN 9780231175272
(pbk. : alk. paper) | ISBN 9780231544375 (e-book)
Subjects: LCSH: Internet—Political aspects. | Information
technology—Social aspects. | Mass media—Political aspects.
Classification: LCC HM851 .R45713 2017 (print) | LCC HM851
(ebook) | DDC 303.48/33—dc23
LC record available at https://lccn.loc.gov/2016041870

∞
Columbia University Press books are printed on permanent and
durable acid-free paper.
Printed in the United States of America

Cover design: Lisa Hamm

CONTENTS

FOREWORD

Culture Industry 2.0, or the End of Digital Utopias in the Era of Participation Culture

ROBERTO SIMANOWSKI

TRANSLATED BY SUSAN H. GILLESPIE

RADIO CAME TOO soon. The society that invented it was by no means sufficiently advanced for it, as Bertolt Brecht observed in a lecture he gave in 1932 on the function of radio: "The public was not waiting for the radio, but rather the radio was waiting for the public." Instead of handing everyone a microphone and bringing society into conversation with itself, Brecht said, people in broadcasting were imitating the old theatrical and print media, addressing the masses from the "stage" of the ether. Brecht thought that the task of turning radio from an "apparatus of distribution" into "the finest possible communications apparatus in public life" was impossible to achieve under the existing social order, but it could be possible in another one, which it was therefore necessary to propagate.[1]

A medium as starting point for the overthrow of an entire social order? The idea isn't so outlandish if we consider the social consequences of the invention of printing. But it would take until the end of the twentieth century before everyone would have access to a microphone. Only with the Internet and then, in real earnest, with the social networks of the Web 2.0 was there a bidirectional medium that allowed every message recipient to turn into a sender. Did that mean that the public Brecht had envisioned for radio was at hand?

This time the medium came too late, although at first people thought its arrival was just in time. That the demise of the socialist social systems occurred in the same year as the birth of the World Wide Web—a historical accident—seemed to argue for the removal of all socially utopian ambitions to the realm of the new media. So it was no surprise when, shortly thereafter, the independence of cyberspace from real-world governments was declared—an idea that admittedly lasted only as long as hardly anyone was actually interested in occupying this space. Today, all the energies of social change are being produced and consumed there, under slogans like "Big Data," "Industry 4.0," and the "Internet of Things."

For a time, the optimism outlived even the commercial takeover that accompanied the new millennium, and it still survives today among some very stubborn types. For the Internet, as cyberspace is now more matter-of-factly known, continues to be a space that is freely accessible: There are no more gatekeepers, no thought police, no elite opinion makers, but instead free access to information and a much-expanded public realm. Admittedly, the often invoked comparison with Jürgen Habermas's study on the historical *Structural Transformation of the Public Sphere* was always already tenuous, since Habermas held that the model of deliberative democracy was better off in the asymmetrical, ideally also self-reflective and multiperspectival discourse culture of traditional mass media than in the symmetrical and decentralized culture of the Internet. For the Internet not only frees public discussion from institutional control; crucially, it also frees it from the central role of political themes and creates a public that is *doubly dispersed*: a public broken down into very small groups, groups that are scarcely willing to consider anything that exceeds the compass of their smartphones.[2]

The "communication apparatus in public life," which for Brecht and numerous others after him promised the emancipation of the individual, undermines—such is the bitter irony of its success—the minimal demand that Brecht posed for radio: that as the locus of political information and discussion it should sharpen society's critical awareness. Brecht's critique of the radio—that "a technical invention with such a natural aptitude for decisive public functions is met by such anxious efforts to maintain *without consequences* the most harmless entertainment possible"—still applies, indeed more emphatically, to the Internet—this despite Wikileaks, political bloggers, and the critical commentary that bravely persists here and there.[3]

The "organization of the excluded," with which radio was supposed to confront the "powers that exclude," has become reality in diverse social networks, but not in order to challenge the status quo, as Brecht and others once hoped. The end of history that was proclaimed in 1989 also spelled the end of Adorno's perspective according to which history—as the emergence of an emancipated, exploitation-free life—had not yet begun. Talk of a life freed from social delusion (*Verblendungszusammenhang*) has faded away. Amusement is no longer disdained as a compromise with false life, and emancipation is now primarily taken to mean self-expression and branding on the social networks. The survival trick of the society Brecht and Adorno wanted to do away with is participation, which, thanks to the combined availability of social and mobile media, keeps people so busy 24/7 that they no longer have any time to think about social alternatives.

The Internet came too late for what it could possibly have become. After the free-market economy, with its tangible consumer culture, had won out over competing contenders for the future of humankind, the new media system could not expect much more from people than what the people had made the new media system into: a virtual shopping center accessible at all times and places, with a few niches consigned to social creativity and political education, niches that, surrounded by advertising and subject to the laws of the attention economy, ultimately serve as supply chains for the neoliberal social model.

But the Internet would not have come at the right time earlier, either. Its apparatuses—hyperreading, multitasking, power browsing, filter bubble, instant gratification, quantification, etc.—are diametrically opposed to the public sphere that Brecht intended. With the next distraction only a click away, patience for things that require effort evaporates. Anyone who doesn't have quick responses to complex questions is promptly and publicly punished by a withdrawal of Likes. But is the medium responsible? Is it the human condition as such? Is the anthropological and technological constellation an overlay over background political and economic interests?

Brecht's contemporaries in the 1920s warned not only about "radiotitis," as excessive radio listening was termed, but also about "tunitis," the constant tuning of the dial from one station to another. The variable tuner was the beginning of the end of the radio as a people's university and apparatus of emancipation, for it made it possible to keep dashing off to wherever something more interesting was going on. No one could have envisioned the temptations offered by its technological successors—the zapping of the remote or the clicking-through of hyperlinks. But the criticism of the radio audience shows that already in Brecht's era not everyone was waiting for the public sphere that radio was beginning to create.[4]

This is approximately what Brecht's media utopia might have looked like from the vantage point of César Rendueles. Brecht's radio theory is an early example of Left media critique, to which Rendueles's book, with its focus on the Internet, is also committed, but without falling into the trap of a utopian extreme. To the contrary, Rendueles's criticism targets "Internet-centrists" and "cyberutopians"—as Evgeny Morozov, in his 2011 book *The Net Delusion: The Dark Side of Internet Freedom*, labeled people whose individual and social lives are completely focused on the Internet or who expect only good things precisely as a result. Rendueles takes Morozov's main arguments against the cyberutopians (data protection, surveillance, slacktivism, excessive faith in technology) and expands them into a principled critique of the illusion that free access to information and interactive forms of communication will automatically lead to a better society. Behind cyberutopianism lies nothing but sociophobia "as a central tenet of economic liberalism," which ultimately leads to a state of modern heteronomy "in which we submit ourselves to the market."

The critique is addressed, on one hand, to the Left, which, starting with Lenin's slogan "socialism is Soviet power plus electrification," usually saw technological progress going hand in hand with modern utopian dreams. On the other hand, it refers to neoliberal capitalism, which takes the Internet activists' slogan "information wants to be free" to mean, above all, "'free' as in 'free market.'" With its "deregulatory strategies"—Uber, Airbnb, and Clickworker.com are worrisome examples—neoliberal capitalism destroys *real* social networks—historically developed solidarities and *Gemeinschaft*, or community, with the root meaning "shared" (*gemein*). This analysis of the media and the overall situation leads to Rendueles's focus—unlike Martin Heidegger but comparable to Bernard Stiegler—on *Sorge*, or caregiving, which should be maintained as the "material basis of our empirical social bonds," beyond social networks like Facebook or Twitter and in opposition to the social destruction of cyberfetishism: "The Internet is great for sharing television series but not for offering care." The ultimate goal of this critique of cyberutopianism, as Rendueles makes clear, is a "reformulation of earlier programs of political transformation" and a "reassertion of their intention to restore social solidarity."

Even if Rendueles does not cite Brecht and Adorno in his argument, the book leaves no doubt that the Internet is finally an ally of the culture industry: "The Internet may be the embodiment of the public sphere, but in that case we would have to accept that the objective of civil society is amateur porn and cat videos. This is not anecdotal. Empirical studies systematically find that the Internet limits cooperation and political critique, rather than stimulating it." That is a sobering diagnosis for all those people who, like Brecht at one time with regard to radio, would like to see the Internet as the "finest possible communications apparatus in public life" and the "organization of the excluded."[5] The method is an old one—one that in the meantime, as Rendueles emphasizes, has increasingly been understood by authoritarian regimes, which are becoming ever more tolerant of the products of the Western entertainment industry: "The Communist Party of China has realized that Lady Gaga is not an enemy but rather an ally." For Adorno, this was the affirmative effect of the culture industry, as one of his most famous sentences reminds us: "The liberation that amusement promises is freedom from thinking as negation." This negation of negation becomes radical when mobile and social media liberate it from specific broadcast schedules and sites of entertainment.

Rendueles's book breathes the spirit of the critical media theory that, in Brecht's and Adorno's sense, has been analyzing the problematic consequences of the new media for ten years or so, with concepts and titles like *Digital Maoism, Computationalism, Alone Together, Networks Without a Cause, The Shallows, Uberworked and Underpaid,* or—perhaps most tellingly—*The Net Delusion.* Some also propose *Reading Marx in the Information Age* and want to combine *Critical Theory and the Digital.* Rendueles's motivation as a writer—as suggested by the symbolic title of the coda, "1989," is the experience of real political energy in the 15-M movement in Spain in 2011, which was precisely not based on the new communication media but became possible only because it was able "to overcome the vicious impasse created by consumerist cyberfetishism." This experience resulted in Rendueles's twofold will both to believe in the possibility of a postcapitalist society and to defend the creation of this society in opposition to the neoliberal mechanisms of the Internet—the social networks, in particular. The result is an essay that reads like an explosive blast. An essay that advances with fulminating speed from one target to the next and flinches from no thesis, however strong. An unspooled hypertext that does not give thinking even a moment to come to rest.

NOTES

1. Bertolt Brecht, "The Radio as an Apparatus of Communication," in *Brecht on Film and Radio,* ed. and trans. Mark Silberman (London: Bloomsbury: 2000), 41, 43. The article appeared originally in *Blätter des Hessischen Landestheaters,* no. 16 (July 1932).

2. Jürgen Habermas, "Political Communication in Media Society: Does D emocracy Still Have an Epistemic Dimension? The Impact of Normative Theory on Empirical Research," in *Europe: The Faltering Project,* trans. Ciaran Cronin (Cambridge, Mass.: Polity, 2009), 138.

3. Brecht, "The Radio as an Apparatus of Communication," 43.

4. Friedrich Pütz, "Die richtige Diät des Hörers" (1927, The proper diet for listeners); and J. M. Verweyen, "Radiotitis! Gedanken zum Radiohören" (1930, Radiotitis! Thoughts on listening to the radio), in *Medientheorie. 1888–1933. Texte und Kommentare*, ed. Albert Kümmel and Petra Löffler (Frankfurt: Suhrkamp, 2002).

5. Brecht, "The Radio as an Apparatus of Communication," 42, 43.

SOCIOPHOBIA

GROUND ZERO
SOCIOPHOBIA

POSTNUCLEAR CAPITALISM

A FATHER AND son walk for days along deserted North American highways that have not seen a car pass in years. Everything is covered by a thick layer of ash, and the clouds releasing their icy sleet barely let a hint of the sun through. The pair's main concerns are finding food and drinkable water, surviving the cold, and not falling ill. They are alone. In this barren land, only degenerate forms of sociality remain. The two sometimes encounter others who, barely human, travel in packs that rob, enslave, rape, and torture. Cannibalism is a constant threat.

This is the plot of Cormac McCarthy's *The Road*, a dystopian novel about a postnuclear future. It may be hard to believe, but many of these things happened, literally and repeatedly, across a vast geographic area over the final third of the nineteenth century. The latter half of the Victorian age was characterized by what the historian Mike Davis, in his remarkable study *Late Victorian Holocausts*, calls a global crisis of subsistence: a holocaust that caused between thirty and fifty million deaths and that is, nonetheless, almost entirely absent from conventional history books.

A huge number of people—predominantly in India, China, and Brazil, though many other regions were also affected—perished from starvation and pandemics during a series of megadroughts, famines, and other natural disasters associated with the El Niño phenomenon.[1] From Kashmir to Shanxi, from Mato Grosso to Ethiopia, the world became a nightmare. Missionaries, one of the usual sources of information about what was going on in remote corners of the planet at the time, recounted terrifying scenes: People used anything and everything as food—leaves, dogs, rats, roofing, handfuls of dirt—before devouring human remains and, eventually, killing their own neighbors to eat them.

Anthropophagy was, in fact, just another step—and not necessarily the last—in the general destruction of the social architecture. Across an immense territory, the rule of law dissolved like an impossible fantasy: Temples were used for kindling, people sold their own relatives off as slaves, banditry was widespread. Over the course of a few years, age-old communal structures disappeared, practically without a trace. Even the physical landscape formed an apocalyptic backdrop: Unprecedented droughts caused the desertification of vast areas, and biblically proportioned plagues of locusts destroyed whatever crops survived. At times, this extreme desertification caused a layer of ash to rain down upon the arid land.

Most of the nineteenth century was relatively peaceful in Europe, at least compared to what had come immediately before. The outlook was not as rosy, however, in the countries colonized by the Western world. From 1885 to 1908, the so-called Congo Free State—the future Democratic Republic of Congo—was quite literally the private property of Leopold II of Belgium, who established there a savage hybrid of turbocommerce, slavery, and brutal violence. More than five million people, and possibly as many as ten, are thought to have lost their lives during those two decades. The Belgian model of commercial exploitation was based on a frenzied rate of extraction that plundered the Congo's natural resources. Leopold II passed a decree that enslaved the entire population, which he then subjected to a reign of terror grounded in mass executions and systematic torture. A common punishment for indolent workers was to cut their hands off and exhibit them as an example.

The ecological hecatombs to which Mike Davis refers were less the direct consequence of colonization than they were the backdrop against which it unfolded and, later, its byproduct. The great powers of the nineteenth century took advantage of the material privations caused by these megacatastrophes to increase the rate and intensity of their imperial expansion dramatically. In most of the world, capitalism arrived quite literally like a military invasion. Humankind had never seen colonization as fast or on such a large scale. Between 1875 and the First World War, one-quarter of the Earth's surface was divided among a handful of European countries, the United States, and Japan. Britain expanded its territories by four million square miles (an area the size of Europe), France by 3.5 million, and Germany by more than one million.[2]

The metropolis developed detailed plans to dismantle local institutions wherever it went. Social structures that had existed for centuries went up in smoke within a few years, in a fairly clumsy and unsystematic—though ultimately effective—attempt to instill a kind of controlled dependence by means of a modern economic, political, and military apparatus. Major ecological catastrophes lent moral weight to this initiative. These countries, cultured Europeans said, were victims of their own backwardness. This paternalistic modernization, painful as it could sometimes be, was for their own good. Karl Marx emphatically voiced this position in an 1853 article titled "The British Rule in India":

> Now, sickening as it must be to human feeling to witness those myriads of industrious patriarchal and inoffensive social organizations disorganized and dissolved into their units . . . we must not forget that these idyllic village-communities, inoffensive though they may appear, had always been the solid foundation of Oriental despotism, that they restrained the human mind within the smallest possible compass, making it the unresisting tool of superstition, enslaving it beneath traditional rules, depriving it of all grandeur and historical energies. . . . England, it is true, in causing a social revolution in Hindostan, was actuated only by the vilest interests, and was stupid in her manner of enforcing them. But that is not the question. The question is, can mankind fulfill its destiny without a fundamental revolution in the social state of Asia? If not, whatever may have been the crimes

of England, she was the unconscious tool of history in bringing about that revolution. Then, whatever bitterness the spectacle of the crumbling of an ancient world may have for our personal feelings, we have the right, in point of history, to exclaim with Goethe: "Should this torture then torment us / Since it brings us greater pleasure? / Were not through the rule of Timur / Souls devoured without measure?"[3]

The reality was much more complex. "Old" has historically been a synonym not of "fragile" but rather of "robust." In the past, traditional institutions found ways to deal, often quite effectively, with the effects of El Niño's megacatastrophes. They created rudimentary support systems that significantly reduced mortality rates or, in the worst case, helped communities rebuild after being ravaged. In contrast, the destruction of their institutional exoskeleton left entire continents socially and materially defenseless. In Davis's words: "Millions died, not outside the 'modern world system,' but in the very process of being forcibly incorporated into its economic and political structures. They died in the golden age of Liberal Capitalism."[4]

The Victorian holocausts established the social order of the world as we know it. They are the model of inequality on a global scale: a relatively limited range of stratification in countries at the center of the world economy (more in the United States and less in Norway, to be clear) and something only vaguely resembling life for one-third of the world's population.

In the West, a combination of institutional arrangements known, not coincidentally, as "social security" offered shelter from the market's volatile climate. The ironic consequence of this is that the center of the "modern world system" refuses to participate in this arrangement to the extent it recommends to the rest of the world. This is a dynamic that can be traced back to Otto von Bismarck but that reached its peak during the Cold War. The founding myth of the so-called welfare states is that they were the result of prudence, consensus, altruism, and lessons learned from past mistakes. In reality, they were part of an intelligent and ambitious strategy led by the United States to minimize the appeal of the Soviet path in Europe. The rest of humanity—that is, most of humanity—was not so lucky. The historical processes that began with the Victorian holocausts created and defined what came to be known as the "Third World."

The consolidation of capitalism on a global scale remains closely tied to momentous destructive processes. The annihilation of traditional institutions configured an ecosystem in which billions of people live. The relationship between constructed space and natural resources on most of the planet is basically what one might expect to see after a megacatastrophe. When Hurricane Katrina passed through Louisiana in 2005, the phrase "Welcome to the Third World" caught on among the hurricane victims of New Orleans. Rather than a tongue-in-cheek slogan, it was in fact an accurate description.

Since the end of the last century, for the first time in history more people live in urban areas than in the countryside. By 2050 the ratio is expected to be 70 percent to 30. Yet it is misleading to talk about a rural exodus to the "city." In fact, there is no agreement among experts regarding the current rate of urbanization because the concept of "city" has been completely blurred. Most living environments today comprise scattered, hyperdegraded settlements that lack the characteristics typically associated with cities. These are conurbations without any discernible outline, without water, electricity, streets, paved roads, or even houses, in the traditional sense of the word. It would be hard to overstate the magnitude of the problem:

> Residents of slums, while only 6 percent of the city population of the developed countries, constitute a staggering 78.2 percent of urbanites in the least-developed countries; this equals fully a third of the global urban population.... The world's highest percentages of slum-dwellers are in Ethiopia (an astonishing 99.4 percent of the urban population), Chad (also 99.4 percent), Afghanistan (98.5 percent), and Nepal (92 percent).... There are probably more than 200,000 slums on earth, ranging in population from a few hundred to more than a million people. The five great metropolises of South Asia (Karachi, Mumbai, Delhi,

Kolkata, and Dhaka) alone contain about 15,000 distinct slum communities whose total population exceeds 20 million.[5]

This is a growing global reality that challenges our perception of social issues. For one thing, contrary to the popular belief that increased life expectancy in the West was the result of sophisticated medical and pharmacological advances, experts agree that the most important factor was the spread of sanitation systems. The most effective weapons against disease invented by humankind are, in fact, cisterns and sewers. On the other hand, the accumulation of excrement in places where these systems are lacking is a major urban issue around the world. Two and a half billion people are literally living in their own shit, without access to any form of sanitary infrastructure, without sewers, cesspits, or latrines. They simply shit and piss wherever they can. This situation reaches Dantesque proportions in places like Kinshasa, a city of ten million inhabitants without any form of waste management. It is estimated that people who live in places without sanitary infrastructure ingest ten grams of fecal matter per day. This is not simply an affront to the senses. More children have died from diarrhea over the past ten years than people have died in armed conflicts since World War II.[6]

Hyperdegraded urban zones, or megaslums, are the colonial problem of the twenty-first century; like the Victorian holocausts, they are the product of economic liberalism. In the 1980s, international economic institutions imposed a program of impoverishment and inequality on the Third World, the true global consequences of which we are only now beginning to understand. Policies promoting devaluation, the privatization of education and health care, the destruction of local industry, the suppression of food subsidies, and the reduction of the public sector not only radically degraded an urban sphere that was already suffering from terrible deficiencies but also sparked a rural exodus that ruined independent farmers while benefiting agriculture and livestock multinationals.

Shantytowns are the other face of casino capitalism, the levees that dam the human remainders of an increasingly speculative and technologized economy. They are a potential source of conflicts on a scale we cannot even imagine. The problems they represent are no longer simply ethical, economic, or political but indeed relate to insurmountable ecological limitations. It is as though the masters of the world were determined to turn Malthus's outrageous nightmares into reality.

The emergence of the Third World has had a profound effect on the political expectations of Western citizens; the reality of a periphery at rock bottom has made us especially sensitive to the cost of social change. The image presented as the flipside of Western liberalism is that of a stupid, irrational, and totalitarian anthropological magma. Deep down, we don't see the alternative to late capitalism as being the solidarity of traditional communities but rather as a hellish cycle of poverty, corruption, crime, fundamentalism, and violence.

In fact, this is essentially the ideological translation of a cognitive bias that psychologists call "loss aversion." One well-known experiment consists of giving various objects to the members of one group and asking how much they would be willing to pay to keep them. The members of a second group are shown the same objects and asked how much they would be willing to pay to buy them. In general, people are willing to pay more to keep something they consider theirs—even if it is something they were given one minute earlier and never wanted before that—than they are to acquire something they do not see as their property, even when the object is exactly the same. From the perspective of rational-choice theory, it is absurd that we should react differently to what is, objectively, the same situation.

Most citizens of Western democracies would pay little to obtain a political system afflicted by a serious crisis of representation and an irrational, unstable, and ineffective set of economic policies, yet they believe the price of getting rid of such a system would be unthinkably high. There may be good reasons to accept things as they are, such as the cost and potential impossibility of a transition to an alternative system, but these are questions we have never even considered. Instead of making rational calculations, we identify change as a terrifying loss. We disparage consumerism, populism, and the finance economy but see them as the last bastion against today's version of the barbarians at the gates. We live in constant fear of anthropological density because the only alternatives to liberal individualism we know are fundamentalism and the squalor of the megaslums. As though there were nothing between the headquarters of Goldman Sachs and the Buenos Aires shantytown known as Villa 31.

Once the ideal of freedom appears, it is there for good, and no political project can ignore it. An anti-Franco activist once told me that when police stormed a student demonstration in the 1960s, he saw one of his fellow protesters try to placate the policeman who was beating him by shouting, "But I don't want freedom! I don't want freedom!" The policeman, probably wisely, doubted his sincerity and went on beating him. When freedom emerges in political life, no one can honestly say they would rather be enslaved. At most we can trick ourselves into seeing submission as a truer form of freedom.

In a similar way, once personal relations of interdependence become the object of suspicion, there is nothing that can redeem them. Like Marx, we can't help but sense something positive in the destruction of communitarian bonds, even if it pains us. The most crude and racist aspect of sociophobia is the fear of a barbarian invasion, of a magma of social holism erupting over our exquisitely and meticulously individualist lives.

At the height of modern colonialism, popular culture reflected these phobias with a sincerity that seems naïve, even amusing, today. The Swiss essayist Sven Lindqvist has collected some fascinating examples from early works of science fiction. In 1910, Jack London, a socialist writer, published "The Unparalleled Invasion." It is a story, set in the future, about the Yellow Peril and demographic crisis. In 1970, China is overpopulated. It is a "fearful tide of life" that has become a monstrous threat of geological proportions: "Now she was spilling over the boundaries of her Empire into the adjacent territories with all the certainty and terrifying slow momentum of a glacier." The West's elegant solution to this Malthusian problem is to exterminate some five hundred million people—according to the text, all the inhabitants of China—with biological weapons and colonize the unpopulated land in order to begin a process of impeccably rational and moderate social reconstruction. Genocide in the service of utopia. In *The Sixth Column*, Robert A. Heinlein's first novel, there was no time to adopt these preventive measures; in Lindqvists's summary, "the pan-Asian hordes have flooded America. The problem is to kill 400 million 'yellow apes' without having to injure real people. The best minds of America hide out in the Rocky Mountains and create a ray that destroys 'Mongolian blood,' but leaves all other blood untouched."[7]

The current version is hardly more subtle, but it is more widespread. To offer one innocuous but meaningful example, the music critic Víctor Lenore explains how specialists systematically trash the Spanish-language music that the lower class listens and dances to as crude, repetitive, and even immoral. Mainstream media push the latest songs from the English-speaking world, despite the fact that their reception in Spain is limited, while it is almost impossible to find news about techno-rumba groups like Camela, whose albums have sold more than seven million copies, mostly among the lower class. Critics view the musical genres favored by immigrants—reggaeton, kuduro, or cumbia, for example—as a bottomless pit of aesthetic degeneration and sexism. It's understandable that fans of abstract music, like the compositions of Karlheinz Stockhausen, let's say, might see contemporary pop as quaint and underdeveloped. This is not the case, however, with most critics, who are always receptive to unoriginal, poorly executed pieces with aspirations to irony—as long as they come preapproved by *New Musical Express*. Most of the music the wealthy West hates is danced to by pairs pressed close together. A reggaeton dance floor epitomizes the symbolic nightmare of the West: a sweaty, poorly educated, tightly packed mass of people singing along to highly sexual and often violent lyrics.

Sociophobia is a universal bias, and there is no escaping it. Many rural and communal movements, nostalgic for the tranquility of traditional social relations and the slow life, are grounded in the image of the big city as a place of social excess, not of individualistic isolation. Walter Benjamin said it perfectly in a text from 1939 called "On Some Motifs in Baudelaire": "Fear, repulsion and horror were the emotions which the big-city crowd aroused in those who first observed it. For Poe, it has something barbaric about it; discipline barely manages to tame it. Later, James Ensor never tired of confronting its discipline with its wildness."

The name of the airplane Ronald Reagan used during his 1980 presidential campaign was *Free Enterprise II*. This bit of caprice turned self-mockery into a kind of marketing strategy. One of the fascinating realities of capitalism is that it has imposed itself on a global scale despite its lack of grand legitimizing discourses. Market society has no Pericles, Cato, or Saint Augustine. There are no declarations of rights, founding charters, or monuments. This is striking because few societies have demanded such heroic loyalty and such an extreme ritualization of daily life. The market insinuates itself into our lives with an intensity that other expansionist, universalist projects—Catholicism and the Roman Empire, for example—never dared to imagine. Yet there is no Arc de Triomphe to commemorate the battles won by the United Fruit Company. No priest says "Abracadabra!" in some dead language to make us accept the transubstantiation of speculative wealth into tangible goods and services.

The dominant discourse regarding our social reality tends to deny it. Politicians only talk about inequality, exploitation, and alienation—objectively, social phenomena characteristic of the modern world—to present them as the collateral damage of an ongoing and inevitable process of improvement. In this sense, radical economic liberalism deserves credit for daring to realistically describe the brutalities of the present . . . in order to defend them. Liberalism has accepted the vertigo of social nihilism. It has embraced sociophobia as a desirable option.

THE GLOBAL PANOPTICON

In Kurt Vonnegut's first novel, *Player Piano*, New York has become something like a private club for the technocratic elite of the United States, which controls an almost entirely automated economy. Most people are shielded from extreme material poverty but live profoundly alienated lives, performing absurd tasks and lacking political agency. At the beginning of the novel we meet the shah of Bratpuhr, an Eastern aristocrat visiting the United States as a guest of the government. The shah demonstrates a great interest in the American way of life. His guide, Halyard, explains the routine of an average citizen: work in exchange for a salary, live in a small house, pay off debts, etc. The visitor, aided by his interpreter Khashdrahr, understands immediately:

> "Aha," said the Shah, nodding, "*Takaru.*"
> "What did he say?"
> "*Takaru,*" said Khashdrahr. "Slave."
> "No *Takaru,*" said Halyard, speaking directly to the Shah.
> "*Ci-ti-zen.*"
> "Ahhhhh," said the Shah. "*Ci-ti-zen.*" He grinned happily.
> "*Takaru—citizen. Citizen—Takaru.*"
> "No *Takaru!*" said Halyard.[8]

From one perspective, our society is quite like all others, and much modern political criticism is dedicated to bringing this to light: the similarities between serfs and salaried workers, the continuity between the slaves that built the pyramids and the children employed by the cotton mills of Victorian Manchester (or, for that matter, the prisoners of Stalin who built the Soviet Union's great hydraulic works). From another point of view, though, everything is different—and these differences are decisive. We have diverged radically from the anthropological norm but have only a faint, nebulous awareness of this essential difference, of its cultural importance, and of our inability to derive any kind of stable system from it.

We have, for the past two centuries, been immersed in an experiment of social engineering on a scale never before imagined. The Hungarian historian Karl Polanyi said that the liberal ideal of a society whose material survival depended on market relations was, simply put, utopian. Throughout history, most communities have used some form of commerce to exchange goods and services, but those traditional markets were always marginal or at least highly limited institutions. The market was a space in the literal sense—the market square—set up at a certain time—market days. Herodotus recounts that when a delegation of Spartans visited the court of Cyrus to warn him of the repercussions he would face if he attacked Greece, the Persian king replied that he was not afraid of a people who had created a place in their cities—the market—for the specific purpose of deceiving one another.

In the modern era the market became, for the first time, an institution that pervaded all aspects of social reality. Commerce has colonized our bodies and souls. We sell off large chunks of our lives in the labor market, we acquire roofs to protect ourselves from the elements by means of complex financial instruments known as mortgages, the air we breathe is traded on the carbon-emissions market, the food we eat is part of a complicated chain of speculation . . .

In contrast, nearly all traditional societies were careful to exclude certain essential goods and services—land, basic sustenance, and money—from the market. Commerce is a form of competition in which each tries to take advantage of an adversary. In the market, "buy low, sell high" is the only unquestionable rule of conduct. Precapitalist societies thought it crazy to tie material survival to the uncertainty of competition for the same reason we think that someone who bets his only house in a poker game or plays Russian roulette is doing something not only risky but in fact misguided: The ratio of risk to reward is simply too high. People will always need food, clothing, shelter, and basic care. Is it reasonable to submit these necessities to the vicissitudes of the market? Does it make sense to close our eyes and hope that the free play of supply and demand will adequately provide for the majority? For millennia, the answer was a resounding no. But, of course, we're much smarter than that now.

The "market system," the phrase Polanyi uses to describe how the market has inserted itself into our lives, resembles a phalanstery or a commune more than it does normal social relations. It is a utopian program and not, as is sometimes said, the peaceable realization of a commercial impulse shared by all humankind. The free market has never existed and will never exist. It is a chimera that has generated unthinkable suffering. And, like all utopias, it is a failed project rife with internal contradictions. Because of this, the state regularly has to intervene in order to keep the capitalist Neverland of the free market from collapsing like a house of cards, taking with it the elites who profit from its false promises. In recent years the same arguments have been employed in defense of the extensive use of public funds to bail out the banking system, the dismantling of government-owned enterprises, and the virtual tax-free status enjoyed by the wealthiest few. Capitalism has never given in to the temptation to be consistent.

Economic liberals call to mind those followers of Henri de Saint-Simon who wore jackets that buttoned up the back so they would need help to close them, thus building a sense of fraternity. The difference, of course, is that the ideology of the market has triumphed and that it now seems like common sense. But we need only dig a bit deeper among the ideological roots of our time to catch a whiff of a pungent, age-old scent that is incompatible with any existing society.

The Yes Men is an art collective that impersonates and parodies representatives of financial institutions and major corporations at international business summits. Their main finding has been that it is impossible to shock the corporate world. Passing themselves off as members of the World Trade Organization, they have publicly proposed initiatives such as illegalizing the siesta, bringing back the slave trade, establishing marketplaces for votes or human rights—so that a country that needs to violate its people's basic rights can purchase an infraction quota from another—and ending world hunger by having the poor "recycle" predigested hamburgers. These proposals were received with interest and murmurs of approval by audiences composed of businessmen and government officials.

Capitalism is impossible to parody. Nothing can shock a world that organizes work, the production of food, and the way money is spent around the omnipresent, compulsory competitive sport we call the market. The worldview of those orderly, sensible, reasonable people who go about their business trying to avoid trouble is an essentially utopian one. Its apocalyptic message has solid philosophical foundations and can be traced back to eighteenth-century utilitarianism. Many critics read the utilitarians condescendingly as naïve pragmatists, petit bourgeois intellectuals without great aspirations. This is a mistake. They are Branch Davidians in disguise. Their ideas seem bland and uninspiring to us simply because their radical nihilist project was done in by its own success.

The founding father of utilitarianism, Jeremy Bentham, was in fact a daring eccentric, a Yes Man of the Enlightenment. In his will, he stipulated that his cadaver should be dissected as part of an anatomy class, then chemically preserved, dressed in his own clothes, and seated in a wooden booth to create what he called an "autoicon." His remains remain at University College London, where they can still be viewed by the public. Bentham dedicated his life to social transformation. He saw himself as a reformer and would not deny himself one last radical postmortem intervention that challenged one of the great anthropological universals: that the appearance of burial ceremonies represents a major milestone in the process of hominization.

Bentham did not reject convention outright. He did not request that his body be tossed in a dump. His body needed first to be treated objectively as dead flesh in order to move on to an improved reformulation of funeral rites. It is a macabre parody of the central element of Bentham's system: the search for the lowest common denominator of sociality from which interpersonal relations could be reconstructed on rational foundations. Bentham recognized the gregarious nature of the human being but distrusted the ethnologically viscous idea of natural fraternity. He aspired to distinguish between sociality and relationships of interdependence, superstitions, unmitigated passions, and false consciousness. He advocated a kind of surgical intervention to correct the communitarian defects of natural social bonds.

At the core of utilitarianism is the idea, not uncommon in the philosophical context of Bentham's day, that all human activity should be judged according to the pleasure or suffering it generates, in the interest of achieving the greatest happiness for the greatest number of people. Bentham turned this cliché into a source of radical political transformation. In essence, the happiest possible group of people is one that allows the individuals that constitute it to perform those activities that bring them the most pleasure, not only in the service of some ethical or ontological individualism but rather as a matter of efficacy: No one, especially not the government, can know what makes an individual happy better than that individual does. The individual pursuit of happiness communicates vital information to the social system, which makes it possible to generate the greatest possible happiness for all. The sources of happiness are specific and diverse; thus, there can be no debate about which aspirations are most desirable.

This strategy is a direct corollary of the idea that pricing is an ideal means of optimizing the distribution of resources. Neoclassical economics was inspired by Bentham. In a perfect world, pricing transmits—at minimal cost—units of information that can be aggregated automatically. In this way, it is believed that a greater level of social coordination is achieved than could be produced by any single institution. From this perspective, centralized planning is bound to disrupt the flow of information, impeding coordination.

For Bentham, maximizing collective happiness is the key to rational social ties. We come together only for economy of scale: Together, we can achieve greater happiness than we could on our own. Any collective intervention meant to structure sociality, including Christian charity, distorts and hinders the individual's quest for satisfaction, which is the only rational motivation that unites us. Fraternity—loyalty, consensus, collective reflection, and reliance on others—destroys the rational basis of society. From then on, this sociophobia has been a central tenet of economic liberalism, though only its most honest, lucid, and morally repellant representatives, like the economist Milton Friedman, would dare to say so:

> To the liberal . . . the ideal is unanimity among responsible individuals achieved on the basis of free and full discussion. . . . From this standpoint, the role of the market . . . is that it permits unanimity without conformity; that it is a system of effectively proportional representation. On the other hand, the characteristic

feature of action through explicitly political channels is that it tends to require or to enforce substantial conformity. . . . Even the use of proportional representation in its explicitly political form does not alter this conclusion. The number of separate groups that can in fact be represented is narrowly limited, enormously so by comparison with the proportional representation of the market. . . . The use of political channels, while inevitable, tends to strain the social cohesion essential for a stable society. . . . The widespread use of the market reduces the strain on the social fabric by rendering conformity unnecessary with respect to any activities it encompasses. The wider the range of activities covered by the market, the fewer are the issues on which explicitly political decisions are required and hence on which it is necessary to achieve agreement.[9]

The market utopia allows us to satisfy our desires without having to face a dense network of religious, domestic, affective, and class connections. It is the difference between simply walking into a store and buying a pair of shoes and trying to get them through the exhausting ritual exchange of gifts known as Christmas. Economic liberals tell us that we are not like the Greeks of Homer's day. In the market we can obtain a three-legged stool, bronze trinkets, and a wineskin without getting caught up in violent competitions, disputes with fickle deities, or exhausting liturgies.

Bentham, however, was much more ambitious. His goal was to extend this project to all coercive aspects of society. The political project of the neoconservative Right in the United States is sometimes mockingly described as right-wing Keynesianism—liberal only in its rhetoric and, in fact, profoundly interventionist. Ever since the Reagan administration, there has been obsessive discussion of the need to limit the influence of the state in favor of the free market. This has been put into practice in areas like health and education—while public spending on the military, police, and the penitentiary system has grown astronomically. Bentham was not guilty of this contradiction. His utopianism was more sincere, and he refused to resign himself to any repression or controls that might diverge from liberal ideals.

The project to which he dedicated the most time, money, and energy was the Panopticon: an architectural and organizational design applicable to any institution where surveillance is required, such as schools, hospitals, military barracks, factories, and—above all—prisons. The Panopticon is a circular structure. The individuals being monitored live alone in cells laid out around the circumference of the building, and the guards occupy an observation tower at its center. Built-in mechanisms such as different elevations, hidden passages, screens, backlighting, and primitive intercom systems allow guards to observe the inmates without being seen.

In the second half of the eighteenth century, when Bentham was writing, debates about the prison system were prominent in the European political agenda. After all, the foundational moment of modern society is marked by the storming of that famous French jail, the Bastille. Enlightenment thinkers wanted to improve the conditions and operation of the prisons. The jails of the time were basically small-scale models of society, chaotic places in which it was literally a challenge to distinguish the criminals from guards and visitors and where inmates' quality of life varied widely depending on their social or economic status. Their daily lives were almost never regulated, and they were often allowed to create their own systems of governance.

Bentham used this microcosm as a laboratory in which to reconstruct social relations on rational rather than communitarian foundations. The technological key to the Panopticon is the permanent visibility of the inmates, who, for their part, never know when they are being observed from the central tower. The uncertainty caused by this exposure generates the same effects as constant supervision, while minimizing cost and personal interactions. In other words, what the Panopticon does is apply liberal sociophobia to the sphere of domination. In Utopia, too, there would be people who force others to do things they would rather not, but the subjected individuals would interact with those subjecting them in an environment free of communal frictions.

The Panopticon is a perfect model for modern international power relations. No one could be naïve enough to think that the relationship between the West and peripheral countries is grounded in magnanimity or that global economic stratification is the outcome of a level commercial playing field. But this domination is subtle and comes cheap to the winners. Like the Panopticon, it does not depend on the intrusive and permanent presence of guards but rather on total exposure to the punishment of the market, international financial institutions, and political accords. Of course, from the Opium Wars to Iraq, no great power has rejected the idea of opening up markets literally at gunpoint, but it is a costly option both politically and economically and might even be considered dishonorable. Surely, Washington has caused more deaths pursuing the commercial interests of the United States than Rome ever did in its imperial expansion, but the prisoners of U.S. wars end up in secret prisons and detention centers, not crucified along Route 66. Real subjugation is reserved only for those nations that dare to break the rules of the international Panopticon, as in the cases of Guatemala, Spain, Chile, Argentina, Brazil, Indonesia, Haiti, Algeria, Nicaragua, and a long list of others that includes Paraguay today.

The free-market utopia has failed. This disaster has led to a succession of increasingly destructive speculative crises, an outcome that is drearily predictable when the quest for private gain wins out over any possible political restriction. An economic system based on an arrogant disdain for the social and material conditions necessary for human subsistence is doomed to fall into a self-destructive pattern that can only end in its attempting in vain to reproduce itself.

The fortunes of Carlos Slim, Amancio Ortega, Bill Gates, and Warren Buffett are purely virtual; they are illusions. No one could convert that much money into currency. Their wealth is, in and of itself, a luxury good. Paleocapitalism was characterized by a naïve culture of ostentation that might seem charming to us today. At one New York dinner party toward the end of the nineteenth century, for example,

> guests found a table heaped with sand and at each place a little gold spade; upon a signal, they were invited to dig in and search for diamonds and other costly glitter buried within. At another

party—possibly the most preposterous ever staged—several dozen horses with padded hooves were led into the ballroom of Sherry's, a vast and esteemed eating establishment, and tethered around the tables so that the guests, dressed as cowboys and cowgirls, could enjoy the novel and sublimely pointless pleasure of dining in a New York ballroom on horseback.[10]

This extravagance, however, pales in comparison to the shocking ambition to accumulate a personal fortune that rivals the GDP of an average-sized country.

The utopia of the Panopticon has also failed. This disaster has led to the creation of the Third World as we know it. The traditional societies of impoverished countries did not simply disappear, giving way to unequal exchange and economic colonialism. Instead, formerly repressed communitarianism has reemerged with terrifying violence. The destruction of such societies has not done away with social friction; it has, rather, transformed it into misery, violence, desperation, fanaticism, and disease:

> In Uganda the Lord's Resistance Army, whose stated goal is to establish government according to the Ten Commandments, recruits members by surrounding a remote school with troops and setting fire to the school. The boys who manage to run out are given the choice of being shot or joining up. Those who join are then required to commit an atrocity in their home district, such as raping an old woman, which makes it harder for the boys to go back home.[11]

A friend from Medellín once told me that the reduction in armed political conflict in Colombia had not reduced urban violence but rather had transformed it. Now murders are carried out by *combos*, teen gangs that fight over the devastated territories of the city's poorest communities. The gang members' taste for rap videos and television from the United States is having deadly consequences. It seems that many innocent bystanders have been killed because the gang members are imitating television gangsters and firing their weapons while holding them sideways, which makes them harder to control and sprays bullets across a large area. Pier Paolo Pasolini believed that consumerism had a destructive effect on society. Today, this is more than just a metaphor.

COUNTERHISTORY

In spite of it all, capitalism has historically been a much more complex and inconsistent reality than we tend to imagine. It is thought that 800 million people around the world participate in cooperatives that employ more than 100 million workers. According to the United Nations, one member of half the households in Finland and one-third of the households in Japan are part of a cooperative. Some 45 percent of Kenya's GDP and 22 percent of New Zealand's can be attributed to cooperatives. In turn, 80 percent of the milk in Norway comes from cooperatives, as does 71 percent of the fish in Korea, 55 percent of the retail market in Singapore, 40 percent of agricultural products in Brazil, and 24 percent of the health sector in Colombia, to name just a few examples. In addition, several million people remain entirely on the margins of the market economy; there are even many who still live by hunting and gathering.

We tend to see multinational corporations as omnipotent, but in fact their power pales in comparison to that of the largest countries. Financial speculation moves astronomical sums of money because it deals in imaginary figures. When it comes to real economics, however, no business comes even close to the income of the world's wealthiest countries. Public-sector jobs represent more than 10 percent of employment, globally—to give one example, the number of individuals who work for Walmart, the world's largest private employer, is only slightly larger than half the number of public employees in Germany. On a global level, domestic subsistence economies continue to be hugely important. Only half of the world's active labor force is engaged in an employer-employee relationship: "Disregarding state capitalism, a major phenomenon in China . . .

no more than 40 percent of the global labour force work directly in a capital-labour nexus."[12]

There are many counterhistories to this modern society of ours that shifts between liberal and panoptic dystopias, and they are not merely remnants of the past to be left by the wayside. In fact, we might find a hidden store of ideas in these experiences that reveal the undiscovered potential of the present moment. Political projects striving for social emancipation are part of this flipside of our time.

Socialist, anarchist, communist, and autonomist movements tried to break down capitalist heteronomy and establish a public space in which it would be possible, at least in theory, to take charge of our lives. Despite what is often said, their project was relentlessly modest. In a poem titled "Communism Is the Middle Term," Bertolt Brecht rejects accusations of radicalism. The radical position is capitalism, which has subverted every material, moral, and ecological boundary. Walter Benjamin completed this idea with his reflections on social revolution: "Marx says that revolutions are the locomotives of world history. But the situation may be quite different. Perhaps revolutions are not the train ride, but the human race grabbing for the emergency brake."

The anticapitalists understood that, in reality, the great dramas of our time—social and material inequality, economic instability, racism, and patriarchal politics—could be resolved with a few minor adjustments: no more than education and the redistribution of the ownership of the means of production. Exaggerating such minor problems to the point of creating a planetary dystopia has doomed modernity to being unable to address much broader issues, such as the sources of personal fulfillment, hatred, and subjugation and the possibility of a nonoppressive model of fraternity. These revolutionaries merely aspired to feed, educate, and bring radical democracy to the world's population, a goal that would seem to be both desirable and possible at our stage of political and technological development. This is precisely what makes the project so alarming: the fact that it could, conversely, suggest that feeding the world's population means destroying the world as we know it.

To be honest, though, these emancipatory programs always retained some aspects of a more ambitious and clearly utopian proposition. Without exception, each encouraged breaking the community-oriented chains of traditional societies that limited individual freedom and extolled authority and superstition. At the same time, however, they denounced modern individualism, the decline of solidarity, and the emergence of social masses held together by extraordinarily weak ties. In this sense they were proposing a return to community, though to a form of community built upon nontraditional foundations. They sought to join the individual liberty that is characteristic of enlightened society with the solid social bonds that contribute to shared personal development. To put it in today's terms, they were trying to propose an alternative not only to the atomized individualism of postmodern consumerism but also to the reactionary return to traditional societies in the form of poverty and fanaticism.

The result, quite frankly, was not very appealing. The socialist New Man aspired to combine bourgeois virtues with robust popular traditions. Relationships of interdependence were traded for indiscriminate solidarity. To judge by Soviet propaganda, the new postcapitalist subject was a heady cocktail of overenthusiasm for great feats of engineering, submission to bureaucratic authority, and a gregarious spirit caught somewhere between a lemming and the captain of a soccer team.

This unique utopian element has been endlessly ridiculed by the same people who, in contrast, speak of the ability of parliaments to embody the will of the people, as though it were some kind of physical variable that could be measured using a represent-o-meter. *The Curious Enlightenment of Professor Caritat* is an amusing utopian novel by Steven Lukes that explores contemporary political theories rather in the manner of Swift or Voltaire. The protagonist, Nicholas Caritat, takes an ill-fated tour through several countries in which the doctrines of communitarianism, liberalism, utilitarianism, and authoritarianism have been taken to their extremes. Significantly, Caritat only visits the socialist utopia, Proletariat, in a dream. This is what a clothing factory is like in Proletariat:

> Nicholas noticed that from time to time workers would rise and
> go from one position to another: a seamstress would join the de-

sign section, a machine tender would become a craftsman, an accountant would take up snipping, and so on. On the highest platform, tall, lithe, elegant, stunningly attractive young women and tanned, muscular and athletic-looking young men walked slowly and sensuously back and forth modeling the clothes made that day. A thousand eyes were continually raised to observe them. In this way, Karl explained to Nicholas, alienation from the product of one's labour was overcome: the workers could, simply by gazing heavenwards, see at any moment the final product of their collective labour.[13]

Though such parodies may be unfair, it is true that the typical conception of the revolutionary social bond is one of the primary and most logical reasons that citizens of contemporary Western democracies reject oppositional politics. Political programs that rely on the emergence of new forms of sociality generate misgivings even among their supporters. It is as if these proposals were not presented entirely in earnest, as though they only exist because those who defend them know they will never have a real chance to put their ideas into practice. It is far from clear why we would stop being individualistic, selfish, suspicious, and unsupportive.

This is not, however, an archaeological dig into the aspirations of oppositional political movements of the twentieth century. Instead, these questions occupy a central place on the horizons of contemporary ideology. Postmodernity has sped the destruction of traditional social ties, throwing the continuity of employment, interpersonal relationships, and political loyalties up in the air. In place of these, it offers an alternative grounded in what one imagines are the new forms of sociality: a growing network of contacts among fragile subjects, dense but tenuous nodes connected with the help of elaborate technological tools.

It is increasingly common for us to refer to interpersonal relationships and group dynamics in terms of the kinds of contact we make online. Political, economic, and demographic changes; cultural output or family ties; emotional or aesthetic experiences— even in places where the Internet and digital devices do not play a significant role, we speak about networks and connections.

We do not feel interpellated by the dual failure of hypercapital-ism and the Third World because our societies see themselves as an ephemeral but solid reticular environment made up of social ties whose profusion compensates for their fragility. In this way, the In-ternet would have made the sociological utopia of communism a reality: a delicate balance between individual freedom and commu-nal kindness—or at least whatever surrogate Facebook and Google+ can provide. Seventeenth-century philosophers used the metaphor of a clock to describe the natural world and human subjectivity. Today, social scientists use the metaphor of the Web to explain all kinds of relationships, whether they are mediated by digital tech-nology or not: immigration, labor, sex, culture, family . . .

I believe that these are fairly weak analogies and that they limit our understanding of long-term historical processes. What is most interesting, though, is how this transformation in the way we think about social relations affects our goal of living in a fairer, less alien-ated world, and what thinking we need to do in order to achieve that aim. Without question, I believe that the fetishization of com-munication networks has profoundly affected our political expec-tations. Simply put, it has lowered them.

Socialism put the establishment of new social relations off for another day: It would be the result of our political imagination and of major social upheaval. Postmodernity tells us that the future is already here, and all we need to do to enjoy it is choose between an Android and an iPhone. What the revolutionary tradition falsely resolved in utopian terms, the geeks falsely resolve in technologi-cal terms. We no longer need Utopia—all we have to do is down-load a torrent client. It is as though the problems of one ideological system were reflected, upside down, in the other. The situation calls to mind Stephen Frears's film *Sammy and Rosie Get Laid*, in which one character defines heterosexual sex as "when the woman tries to come, and can't, and the man tries not to, and can't."

For emancipatory traditions, fraternity would be the result of slowly and laboriously overcoming some of the material, social, and political problems of modernity. Contemporary futurism inverts this formula: The digital revolution aspires to do away with the eco-nomic problems of the free market by favoring new commercial relations based on knowledge, creativity, and connectivity. It would also erase the disaster of the global Panopticon with a single stroke. Underdeveloped countries would break their cycles of abject poverty and commercial dependency. Today, many Africans use advanced mobile devices without ever having owned a computer. In this way, the most disadvantaged countries will skip entire stages of development and gain access to a conflict-free economy without ever having to pass through industrial purgatory. India would go directly from being a land of dispossessed peasants still restricted by the caste system to an egalitarian society of programmers, engineers, hackers, and community managers. Egypt, from a Third World dictatorship protected by the West to the most advanced cyberdemocracy . . . And all this without guillotines, winter palaces, wartime economies, import substitutions, literacy missions, or vaccination campaigns. All this by simply embracing not just the market, now, but its new and improved version: digital interactivity.

I believe that cyberutopianism is essentially a form of self-delusion. It keeps us from seeing that the primary limitations of solidarity and fraternity are inequality and commoditization. However, I also realize that the classical emancipatory projects—socialism, communism, and anarchism—have failed, at least in the strict sense of their programs. Not because their demands are unreasonable in today's context, or because they have been met. In fact, the opposite is true. The problem is that freedom and equality are too urgent and important to be left to programs with which so few people identify. A society that thinks of itself as a network is not the same as one that does not. Therefore, any critique of cyberutopianism should lead to a reformulation of earlier programs of political transformation and to a reassertion of their intention to restore social solidarity.

Deep down, the social effervescence of digital media is inessential, decorative. It is useless at fostering what our shared existence should: taking care of one another. The same goes for egalitarianism 2.0, the feeling that social difference is minimized on the Web. Radical democracy is not a universal customer-service line. It's a little crazy, if you stop to think about it. It means that the idiot with the Porsche Cayenne, or the lady who lets her pit bull loose in a park full of children, or the bridge-and-tunnel types at the mall have the same right to intervene in public life as you. The historic Left knew how to present this outrageous idea in a way that made it seem both possible and desirable to most. I don't think this program can be revived, but, of course, we must replace it with ambitious antielitist projects that also openly address the sociological dead end in which the Left has ended up: the search for a cohesive and viable social order that is also compatible with individual autonomy and personal development.

In short, I believe that complex, enlightened societies possess the raw materials necessary to tackle the issues of democratization, equality, freedom, and solidarity without falling into the trap of reactionary collectivism or the chimera of the socialist New Man. The ideological fetishization of the Web, however, is a major obstacle in this pursuit.

In this book I will first explore contemporary cyberutopianism, paying special attention to those elements that are considered most politically progressive, and will then take a backward leap to examine issues that traditional anticapitalist programs have failed to address. My aim is to cause a kind of shock so as to defetishize the current futurist ideology and allow the emergence of previous possibilities that have been buried until now. Perhaps some practical use will come of this.

At the very least, juxtaposing communist utopianism and the ideology of a reticular community will help us grasp something about the nature of social ties in the postmodern era. The Internet, I argue, is not a sophisticated laboratory in which delicate strains of the communities of the future are being developed but is instead a rundown zoo housing the decrepit forms of age-old problems that still haunt us, though we prefer not to see them.

1
DIGITAL UTOPIA

CYBERFETISHISM

TECHNOLOGICAL DETERMINISM, PARTICULARLY in its Marxist form, has gotten a bad rap—at least if the technology in question drips oil, belches smoke, weighs a ton, or is, generally speaking, analog. For a long time, those explanations of social change that considered applied science to be a critical factor were deemed unsophisticated and monocausal (a bad thing, it would seem). Today, technological determinism is back with a vengeance, but only in relation to information and communication technologies. No one is interested in arguing that advances in turbo-injection technology produce meaningful social transformations—though they almost certainly do. In contrast, judging by its impact on the media, an update to Twitter's timeline is received as a social change as fundamental as the Neolithic Revolution. The only solution our governments offer to the economic abyss we face is to repeat the mantra of the "knowledge economy," a panacea for everything from structural unemployment to world hunger and pollution.

In fact, a certain degree of technological determinism is not only plausible but inevitable—at least for those of us who think that the social sciences should aim to discover the causes behind observable phenomena and not only provide literary interpretations of these. The reality is that the term "cause" is used much more loosely in sociology or history than it is in the natural sciences, where it is essentially synonymous with universal and quantifiable regularities.

The physical sciences have lodged an idea of causation in our heads, namely that causes are the catalysts of effects that can be verified with precision: typically, one body acting on another and causing a change in its path. But history and the social sciences work with models of causality that are not so much complex as they are confusing—just as happens in our own lives, in which we are unable to formulate exhaustive explanations for our experiences. In our daily cognitive practice, we often identify those enduring systems that offer the most resistance to change as causes.

Causes, in this broad sense, limit the horizon of possibilities rather than bring about a clearly defined effect. We tend to identify causality with the ability of a system of events—or something that we understand as one—to resist change. For example, when we say that the education a person has received has a significant influence on the way he or she is, we are not identifying a clear causal chain but are rather indicating a set of habits that parents transmit to their children and that persist throughout the various stages of their lives. In the same way, identifying the causes of the financial crisis entails pointing to how it came about despite the best efforts of many individuals and institutions to prevent it.

Applied science is, in principle, a reasonable place to look for these kinds of causes. The technology available to us conditions the way we relate to our surroundings and the way our society is organized. Technology is also less easily influenced by social change than other phenomena. Though in theory one could point to countless nuances—constructivists specialize in precisely this—it seems reasonable that changing the legislation governing factories that produce combustion engines is easier than transforming the combustion engine itself.

Nevertheless, this form of causal attribution based on longevity does not offer in and of itself any information about how technology influences social relations, if indeed it does, in any more than a very general way. We are fairly certain that a given society's level of technological development is closely related to enduring social structures. For example, slavery did not play an important role in hunter-gatherer societies. The reason for this is not that pre-Neolithic societies were intrinsically kindhearted but rather that contexts of low technological development produce little surplus. All members of a community would have had to work to ensure its survival. As such, slaves would not lighten their masters' workloads; they would, in fact be a burden on the available natural resources.

In general, there are reasons to believe that technological advances are positively correlated with increased material inequality throughout history. But assertions of this kind are extremely vague, almost clichés. In the 1950s, the economist Simon Kuznets tried to turn them into a sophisticated, empirically well-grounded theory. Decades of increasingly complex tests meant to prove this claim have produced shockingly poor results: Technological advances are compatible with greater equality in those societies that are committed to the redistribution of wealth and egalitarian ideals.

A completely different and far more concrete matter is what technology can offer politically. Technological progress has gone hand in hand with modern utopian dreams. When Lenin said that socialism was the Soviets plus electricity, he was expressing a deeply rooted belief held not only by the political Left. In the 1930s, Le Corbusier proposed demolishing the entire historical center of Paris, just a few decades after Baron Haussmann had done just that. His argument was both technical and poetic: "In order to create the organic architectural entities of modern times, the soil must be redivided, made free and available. Available for the undertaking of the great works of machine civilisation."[1]

Through numerous schools and reformulations, this philosophy has come to pervade the ideology of contemporary architecture. Many architects feel authorized to practice a form of social engineering that is as naïve as it is ineffective, sometimes in a friendly, well-meaning way—adapting their work to local communities, at least, to how they imagine them from their cantilever chairs—and at other times aggressively, trying to force large-scale social change. Lewis Mumford summed up the limitations of this approach quite well, writing that "the gains in technics are never registered automatically in society: they require equally adroit inventions and adaptations in politics; and the careless habit of attributing to mechanical improvements a direct role as instruments of culture and civilization puts a demand upon the machine to which it cannot respond."[2]

Marx's position in this regard was fairly complex and not without contradictions. As we know, for Marx, technology played an important part in historical change. Nonetheless, when it came to socialist emancipation, technology's role was purely preparatory.

Marx's thesis is, in fact, fairly pessimistic: Without meaningful material advances, it is impossible even to imagine political emancipation. As long as scarcity prevails, cooperation and altruism do not stand a chance. Socialism requires a context of material abundance; this is precisely the situation in which the Industrial Revolution began. Capitalism is a sort of window of opportunity for emancipation, one we must take advantage of before it self-destructs. The idea is that, once the forces of production have reached a certain level of development, the political decision to make efficient and egalitarian use of technology could put an end to Hobbesian confrontations and open a new space for friendlier political relations. Social revolution is the process of making this decision. On the other hand, Marx did not expect technology to play any particular role in promoting emancipatory social relations or overcoming alienation once this new context of autonomy had been reached.

Contemporary technological determinism proposes exactly the opposite. In the first place, it does not contend that major political change is necessary to maximize the social usefulness of technology. On the contrary, according to this way of thinking, contemporary technology is postpolitical in the sense that it exceeds the traditional mechanisms that bring order to the public sphere. Second, technological determinism sees technology as an automatic source of liberating social transformation. For this reason, rather than technological determinism, one would have to talk about technological fetishism, or—given that most of this ideology unfolds in the realm of communication technology—cyberfetishism.

The expression "commodity fetishism" appears in a brief passage in the first volume of *Capital*. Marx uses it to describe how, under capitalism, the nature of certain major social processes only reveals itself through market effects, leading us to view what really are relations between people as the commercial interactions of goods and services. As part of the market, we understand one another through the goods we buy and sell. This is precisely what supporters of the Californian creed argue from their vast, Internet-centrist party headquarters in Silicon Valley. From their point of view, not only is the interaction among devices laying the foundations of a more fair and prosperous social order; it is actually bringing about that social change.

Cyberfetishists claim that technology is vitally important—though, judging from their arguments, its influence seems to be exerted magically. These cyberfetishists offer no concrete clues about how technological change affects social structures. Because of this, their proposals tend to be either highly ideological—sometimes explicitly so, taking the form of a manifesto—or very formal, focused on ethical or legal questions rather than power relations and the material conditions into which they insert themselves. No one would have imagined thirty years ago that a few Harvard professors would become icons of protest movements and engaged citizenship around the world.

To be fair, in recent decades copyright issues have proven central to conflicts that directly affect the economy, international relations, access to public resources, and civil freedoms. The reality is more complex than the theoreticians of cognitive capitalism would have us believe. There is surely some kind of conceptual connection between Monsanto's biopiracy and the lobbyists that fight to keep Hollywood movies out of the public domain. But a rural community in Kerala and a fan of old movies from North America face realities that are radically different, in ways that concepts like "general intellect"—a notion Marx presents in the *Grundrisse*—completely fail to capture.

It is also true that until very recently, patents and copyrights were part of an obscure and fairly dull area of commercial law. Even back then, however, word of scandals involving intellectual property rights, such as the mass confiscation of pirated music by British police, would occasionally reach the media. The matter was also, of course, of great interest to businesses and governments. In fact, legislation and the commercial strategies that developed in relation to intellectual property played a significant role in several of the courtroom battles that shaped the past century's industrial monopolies and international relations.

For example, at the beginning of the twentieth century, when the United States had already become the leading industrial world power, Germany remained at the forefront of the strategic field of applied chemistry. In 1912, 98 percent of all chemistry patents issued in the United States were held by German companies. Things changed during the First World War. According to David Noble, "The war, with its unprecedented need for organic-based explosives, and thus a domestic industry independent of Germany, changed this situation dramatically. The U.S. government . . . seized all German-owned patents . . . [and established] a private foundation to hold the patents in trust and issue licenses to American companies on a non-exclusive basis."[3] Between 1917 and 1926, more than seven hundred confiscated patents were issued to American businesses, putting those already powerful organizations in a dominant position. Among the companies that benefited most from the expropriated patents were names like Du Pont, Kodak, Union Carbide, General Chemical, and Bakelite.

Despite their importance, however, these processes never had the economic impact, public visibility, and political significance that they do today. Just a few years ago, it would have seemed absurd to us that a large-scale FBI operation against the New Zealand–based company of an eccentric German millionaire would make the front page of newspapers around the world and would be a source of concern for thousands.

Some of the most influential technology experts of our time deal with questions of intellectual property. Legal issues are at the core of recent technical and scientific discussions, edging out interest in the effects that technology has on social structures, power relations, and individual identity. In this context, the most forceful popular voices have aligned themselves with the free circulation of information and against the copyright industry.

The corporate world has lost in the court of public opinion. Julian Assange graced the cover of *Rolling Stone*, Lawrence Lessig appeared on *The West Wing*, Justin Timberlake played Sean Parker in *The Social Network*, Linus Torvalds inspired characters in Hollywood megaproductions and has a meteorite named after him, and Richard Stallman has become a countercultural icon. The industry has achieved noticeably poorer results in the cultivation of its image. In the *South Park* movie, a general executes Bill Gates when Windows 98 freezes on him, and in a recent episode of the series, Steve Jobs is presented as a Josef Mengele for the digital age.

Disputes over copyright are spilling over into the debates of social movements in the analog world. In fact, one of the catalysts of the 15-M antiausterity movement begun in May 2011 in Spain was the campaign against the *Ley Sinde*, or Sustainable Economy Act, which was aimed at limiting downloads of copyrighted material from the Internet. Reflections on common property and its relationship to the market can be traced back at least as far as Marx's early writings in the *Rheinische Zeitung* about legislation regarding the theft of firewood. Only recently, however, have such reflections come to play a central role in explaining the fundamental dynamics of capitalism and its alternatives. Copyleft initiatives have called attention to the expropriation of common property as a systemic presence not only during the heroic age of industrialism but in today's economies, as well.

I don't think it would be an overstatement to say that movements favoring open access to knowledge incorporate modified strategies from the Left aimed at curbing the neoliberal counterrevolution. This is especially ironic because many intellectual-property initiatives have little to do with programs of political liberation. Some of their protagonists, in fact, are right at home in a commoditized, classist environment.

The reason these copywars are of such interest to activists is that they consolidate many of the issues identified by anticapitalists over the past two centuries. We live in a uniquely paradoxical economic system: It creates incredible technological and social opportunities on which it is often unable to capitalize. Modern society has specialized in turning what should—intuitively, at least—be solutions into systemic problems. Technological advances generate strikes and unemployment, rather than free time; increased productivity leads to crises of overaccumulation, rather than abundance; the media create alienation among the masses, rather than enlightenment . . .

Looking at copyright law, we begin to see not only that today's societies tend to privatize gains and socialize losses but also that they have trouble dealing with material abundance when the distribution of this is not commoditized. The thought of doing away with the labor market, understandably, makes many people's heads spin. They believe that there is something in the nature of things that makes the competitive relations of the market an inevitable, or even desirable, way of dividing labor in a complex society.

From the perspective of the average economist, in market-driven societies there are not only moral but also causal connections between the quest for personal gain and the way that the supply of goods and services is organized. Bakers would have no motivation to show up to work every morning if they made no money by doing so—and neither would those who produce flour or those who harvest the wheat. What is more, they would have a very hard time figuring out how much bread they need to make, what kinds, how much flour they need, and so on.

When it comes to intellectual property, the organizational merit of the market in the context of an abundance of digital material is much less clear these days. There are some who believe that if rock musicians lost all hope of becoming multimillionaires, they would burn their guitars on a pyre. This is basically like thinking that if the state lottery were to disappear, we would all fall into an abyss of desperation at the thought of living out our lives in material mediocrity. Still, regardless of whether the market incentivizes creativity, there is no denying that the only deterrents keeping a finalized digital file with no expiration date from being distributed ad infinitum nearly free of cost are social rather than material. This is not the case with most of the goods and services produced within the market.

With digital goods, the relationship between supply and demand is much more complex than it is in the average commercial context. On the one hand, abundance only exists for past production: Present and future production are still costly and rare. There are artists who expect to be compensated or financed and who cannot or will not offer their products under other conditions. On the other hand, symbolic complexities thrive when price is not a barrier to distributing goods that have already been created, transforming the relationship between what people expect and what artists can and will provide. Aesthetic, emotional, and political factors permeate the relationship between supply and demand with an intensity that is inconceivable in the context of the market. These factors influence artists, leading them to take on projects they would not, for free or for a fee, in the traditional market. From a conventional economic perspective, dedicating a huge amount of time and energy to, say, subtitling an obscure Japanese animation series anonymously and for free is almost irrational.

For this reason, issues surrounding copyright also have a propositional dimension. In the first place, many people sense that the copyright wars contain the seed of an alternative to the dead-end street of 1970s Keynesian economics, a third way between state bureaucracy and privatization. Programs critical of the copyright industry often develop novel cooperative strategies. There are many initiatives with a strong altruistic component and little centralization that foster new forms of coordination. What is more, many of these have no commercial mandate and receive no support from formal institutions.

Second, the debate surrounding copyright seems to be developing in an ecumenical space particularly well suited to allowing the Left to overcome its organizational difficulties. The conflict over intellectual property appears to bring individuals from diverse ideological traditions into agreement. At the same time, however, the points of consensus—decommoditization, altruism, reciprocity—bear a strong family resemblance to the classic program of the Left.

At least since the *Communist Manifesto*, anticapitalism has aspired to universality. The socialist project centered on the working class, but only as insofar as it was representative of basic human aspirations. With the cooperative movements of the Internet, the Left seems to have found a cool, technologically advanced version of its old universalist tradition. Today, playing the role of the consciousness of liberation in place of the sans-culottes are the sans-iPhone, who participate in cooperative digital projects as members of an enlightened and socially engaged avant-garde. For the first time in a long while, activists are sharing platforms and projects with individuals outside their organizing traditions and even with those who hold opposing positions. Jimmy Wales, the founder of Wikipedia, is an anarchocapitalist who cites Friedrich Hayek just as freely and well as he does the famous hacker Erik S. Raymond. The underlying cause is that the Internet is increasingly seen as the fullest realization of Habermas's ideal of communicative action: free individuals interacting without any analog impediments, in such a way that their shared rationality can emerge unhindered.

I believe that both these ideas, copyright and technologically mediated universality, are essentially erroneous. Copyright is, without a doubt, a political battlefield, but in no way does it offer an automatic solution to the practical dilemmas we have inherited. Instead, it reproduces these in the sphere of communication networks, where a combination of utopianism and fetishism tends to render them invisible.

Hopes for social development based on one technological innovation or another have been repeatedly dashed given the need to overcome restrictions stemming from both the market and the actions of the state. One notable example is the project founded by Nicholas Negroponte to fabricate hundred-dollar computers, the results of which were profoundly limited by a paradigmatic combination of commercial and institutional obstacles. The initiative, known as One Laptop per Child (OLPC), sought to mass produce low-cost portable computers specifically designed to be used by children in poor countries.

The early stages seemed promising. In general terms, the prototypes were well received by specialists in the field. The problems began after Negroponte found a manufacturer in Shanghai willing to produce the computer. This company invested funds to accommodate the anticipated initial orders: some seven million in the first year. However, the actual orders placed were only for one million devices. The manufacturer recouped its costs by significantly raising the price of the computers it had already produced. On the other side of things, OLPC was not able to find trustworthy institutional channels—governments and educational organizations—that would buy and distribute the computers through appropriate public programs.

In summary, it was physically possible to manufacture the hundred-dollar computer, but not within standard commercial structures. In fact, it is widely accepted that the boom in netbooks and tablets was a direct result of the OLPC project, which identified an overlooked niche in the market. Netbook manufacturers simply eliminated all social and educational considerations from the project and interpreted in strictly commercial terms its goal of producing a computer that was—though not very powerful—small, cheap, and independent. Additionally, it quickly became evident that the OLPC project could only be readily implemented in wealthy countries with established educational systems, where it really wasn't necessary, or in those few poor countries with solid institutional foundations. It is no coincidence that one of the places where OLPC has had the greatest effect has been Uruguay, a country with a leftist government, one of the best educational programs in the region, and a literacy rate of nearly 100 percent.

The prevalent image of the Internet as a privileged platform for fostering democracy, participation, and cooperation has repeatedly run up against reality in just this way. The media and specialists in telecommunications are inclined to bend the facts as far as necessary to avoid any sort of political opposition to the byproduct of communication technologies. The truth is that not only does open access to the Internet not lead directly to critical thought about politics and civic engagement—if anything, it stifles them.

In his rigorous study, Evgeny Morozov analyzes—among many others—the case of Psiphon, a copyleft tool developed by the University of Toronto's Citizen Lab to facilitate anonymous access to the Web by citizens of countries affected by Internet censorship.[4] Psiphon turns the computers of users collaborating from countries without censorship into proxy servers to which users living in countries where the government controls communication technologies can connect. A secure, encrypted connection is established between the user and the Psiphon proxy that cannot be intercepted. In other words, it is not a centralized solution to censorship but rather a distributed, collaborative, copyleft network. It would seem to be cybernetic Utopia itself. Yet the Western participants in Psiphon discovered that a large number of individuals who requested access to Psiphon's servers from China and other countries that practice censorship were spending their time online looking at pornography and celebrity gossip, not downloading information about Amnesty International. The Internet may be the embodiment of the public sphere, but in that case we would have to accept that the objective of civil society is amateur porn and cat videos. This is not anecdotal. Empirical studies systematically find that the Internet limits cooperation and political critique rather than stimulating it.

⊚ ⊚ ⊚

Some time ago, the satirical newspaper *The Onion* published an article with the title "Drugs Are Winning the War on Drugs." Something similar is happening with the copyright industry's attempts to protect its monopoly. Shutting down websites from Napster to Megaupload, the ongoing World Wide War presents, in radical terms, an economic problem that exists in both historical materialism and theories of creative destruction going back to Joseph Schumpeter. The capitalist economy maintains a paradoxical relationship with technological development. Innovation is a key source of earnings, but at the same time, it is detrimental to consolidated sources of surplus value.

The digital revolution is a perfect example. Essentially, it has had two irreconcilable consequences. On one hand, the release of masters has turned artistic and cultural products into common property, in the sense that economists apply to the term. On the other hand, digitization has multiplied the chances of benefiting from intellectual property at little cost. Beyond a certain threshold, copyright is a source of speculative earnings with only a distant relationship to actual production.

Public goods are not necessarily those administered by the state. They are characterized by the fact that the access of people who already use them is not limited by the appearance of new users (in Economese, they are "nonrivalrous"). Another important characteristic of public goods is that their use cannot be limited by market-based mechanisms (as such, they are "nonexcludable"). Anyone can enjoy them, whether or not they have contributed to their production; as such, the cost of producing these goods cannot be covered by putting a price on them.

Public goods and intellectual property have always coexisted in a state of unstable equilibrium. Analog radio and television broadcasts were public goods provided by state and private entities. There was no way to limit access to them technologically: Anyone with an antenna could tune in to the signal without exhausting it. It was impossible, or very difficult, to make users pay for the content they consumed. Much like the performance of a street musician, any passerby can enjoy the music, and the artist cannot limit access to his or her creations by charging for tickets.

Copyrighted products, on the other hand, have very different characteristics. Analog books and recordings are, typically, rivalrous and excludable. If I am reading a copy of *The Brothers Karamazov*, it is unlikely that you would be able to use the same volume simultaneously. To acquire the book, we have to go to a store, which limits our access through the imposition of a price (we could also go to a library, but that is another matter). Nonetheless, the notion of intellectual property raises questions even in these cases: Records and novels are not public goods, but what about a poem or a melody, either of which someone with the necessary skill could commit to memory and repeat?

These questions have no simple answer. It is difficult to establish clear borders in the shifting landscape of immaterial production. This is why intellectual-property legislation is so dense with stock phrases and has such an air of artificiality. What gave it meaning and made its rules more or less acceptable was its objective, which was much more intuitive. It sought a system of legal counterbalances that could accommodate the interests of artists and authors, middlemen, and the public. This meant, essentially, allowing creators and producers a monopoly over their work, but a limited monopoly created with the general public in mind.

The legislation governing intellectual property in the West was marked by the decision to leave a large part of the work of producing and distributing immaterial goods, and of compensating artists and authors, to the market. The results have been mixed. The cultural output of the past century has been immense, at least in quantitative terms. The price we have paid has been not only the commoditization of this output but also the proliferation of biases related to class, gender, and race. For decades, the world has been living under the influence of a startling Anglo-Saxon cultural hegemony. And that's to say nothing of the ideological filters at work in disseminating information.

Choosing the market system had much more to do with protecting the middleman and private distribution—that is, with a commitment to the copyright industry—than with any attempt to incentivize creativity. It was a deliberate choice; there were other options. After all, noncommercial backing has historically had some fairly decent results—Greek tragedy and Renaissance art, for starters.

The culture market is hardly hegemonic these days, either. Highbrow music, for example, has almost always been supported by organizations with noncommercial objectives. In the publishing world, nonprofits often foster the publication of genres that are well regarded but less commercially viable, like essays and poetry. In certain countries, public television is financed by the direct taxation of those who use them. Some museums receive funds through donations; similarly, street musicians hold out their hats to passersby. It is quite possible to imagine that a different system of production, distribution, and compensation might have developed, one in which the market played a marginal role, or at least not such a central one. One example along these lines is that of basic science, which is supported by a wide range of public and private institutions: universities, research centers, armies, foundations, businesses . . .

In any event, the traditional system of remuneration for creating music or film based on reproduction and distribution has collapsed with the processes of digitization and the popularization of the Internet. The rise of the e-reader predicts a similar future for the publishing industry and print media. The dominant payment models are used by creative professionals who are able to keep their output from becoming common property through the control of hardware. This is the case with video games and stage acting. Other models of financing—for example, those based on voluntary micro-donations—are theoretically possible but still relatively marginal and effective only in limited cases.

Ironically, the crisis in the traditional distribution and remuneration systems for intellectual property has run parallel to exponential growth in the gains derived from the copyright industry and their impact on the countries at the center of the global economy. In recent decades, intellectual property has become a key component of capitalism.[5] The three sectors that generate the greatest revenue for the United States—the chemical, entertainment, and software industries—are all based on some form of intellectual-property protections. Emphasis is usually placed on the relationship between intellectual property and technological innovation and its effect on economic growth. In contrast, the organic relationship between copyright and unproductive gains is almost never discussed. The same technologies that convert certain forms of intellectual property into common property can also turn it into a source of speculative profit.

In the canonical, respectable version of capitalism, financial products are created to anticipate future productive initiatives and inject liquidity into the economy. The monopoly afforded authors, artists, and distributors over their intellectual property is analogous to this. It guarantees that a creative investment—consisting of time, effort, talent, and money—will not be undercut by parasitism. In both cases, the reality of Western economies has for some time now been turning the original terms of the social contract upside down. According to the Bank for International Settlements, in 2007 the sum total exchanged in financial transactions was seventy times the world's GDP. Speculation is the chief source of profit in contemporary Western capitalism; at the same time, the right to monopoly has uncoupled copyright from its original objectives and made it an end in and of itself.

Obviously, not all ties between the real economy and the world of finance have been broken. Goldman Sachs, for example, reaps astronomical gains from its speculation in agricultural derivatives. These investments are possible because of the existence of a large-scale agricultural industry and demand to match. Similarly, successful intellectual products are a necessary component of the copyright industry in the digital era. For the moment, there are no secondary culture markets (though in 1997 David Bowie did list the rights to his music on the stock exchange). The real source of copyright-based profit is the technological, commercial, and cultural ability to sell commodities that, above a certain threshold, are produced with barely any marginal cost. The great intellectual-property tycoons can derive gains practically without any investment at all in production. Like speculative finance, the copyright industry is a press for printing money, and we all pay the price for this privilege.

Some of the reasons why we tolerate this strange situation are ideological. We tend to consider extreme finance capitalism and the most speculative practices of the copyright industry to be aberrations that stand out against the legitimizing normality of the knowledge society. Many people—including more than a few heterodox social scientists—are convinced that in the present economy value creation is based on immaterial cognitive practices and that this implies a major break from any situation in the past.

The very notions of immaterial labor and a knowledge economy are confusing. They bring together vastly different processes under the same label. Developing software may require creativity, though perhaps not more so than, for example, engineering did in the early twentieth century. On the other hand, though the labor of a telemarketer is equally immaterial, it bears a far closer resemblance to activities typically associated with Ford's assembly line. The truth is that information technology, just like the old industrial machinery, can enhance or reduce the worker's capabilities. Some fast-food multinationals use terminals with symbols and icons that make it unnecessary for their operators to know how to read or write.

The challenge of dealing economically with the elusive nature of creative labor has historically led to the search for compromises that allow it to be remunerated and safeguarded without getting tangled up in futile digressions regarding the precise nature of intellectual production. For example, because it is so difficult to anticipate which scientific research will be fruitful, one strategy for protecting such research has been to join it to teaching at universities. We pay university professors for their visible and controllable labor—teaching—and allow them to dedicate another part of their time much more freely to research. Something similar goes on in the economy as a whole. There is no question that intellectual labor, in a broad sense, is centrally important; its centrality may explain, in part, the distribution of profit in today's economies. What is much less clear, however, is exactly how intellectual labor is the source of this profit, beyond the trivial fact that research and invention are sometimes required in order to bring new, competitive products to market.

The geographic distribution of highly qualified immaterial labor explains why profits from the sale of iPads are concentrated in a few businesses in California rather than in the Chinese factories where the devices are assembled. From another point of view, however, the importance of knowledge in some of the most lucrative fields is conditioned by processes rife with conflict. In recent decades, the number of unskilled manufacturing jobs has not diminished but has instead greatly increased on a global scale. This explains, for example, how iPads can be produced on the cheap and, as a result, be sold on a massive scale. As Erik S. Reinert says,

Nations specialized in the *production* of new technologies generally experience very different effects from the consuming nations or the nations supplying the raw materials needed for that same technology. . . . For example, information technology (IT) creates very different results around Microsoft's headquarters in Seattle compared to the hotel industry. In the hotel business as well as in the used book business across Europe, the use of IT has led to falling margins and increased downward pressure on wages and profits.[6]

On the other hand, it is impossible to establish a clear distinction between immaterial creative labor and the parasitic version of it, which resembles speculation. At one extreme lies the invention of a vaccine for an untreatable disease and at the other, biopiracy; between these, however, extends a vast repertoire of more ambiguous practices, like the development of technologies with aggressive barriers to access.

Put another way, it is impossible to isolate the central role of knowledge in contemporary value chains from the division of labor in the context of international competition. Global inequality is not an intrinsic consequence of the relationship between techno-science and the market economy. The factor that determines who gains what in the global knowledge economy is class struggle, not the product of some blind study in the journal *Nature*. Theorists of the knowledge society give the impression that they are analyzing some kind of natural tendency in the most successful capitalist societies toward heavenly immateriality. In reality, what we get is a biased description of the political, economic, and even military strategies developed by countries at the center of the world economy to keep those on the periphery under their power.

From the 1970s on, the wealthiest countries have tried to lay claim, simultaneously, to those production processes with the highest value added in order to drive up their speculative gains. The protection of intellectual property combines these two dynamics through the law. The industries that generate the highest profits depend on safeguards to protect intellectual property, and governments are more than willing to provide this legal protection. At the same time, these businesses systematically use their technological advantage to speculative ends. Monsanto has the resources and the technology to undertake biological research and uses the protection afforded it for that research as a cover for biopiracy. Hollywood is in a position to inundate the rest of the world with its product, which is why it wants to make sure its movies do not cross over into the public domain. Microsoft and Apple (and, on a smaller scale, Oracle and Adobe) have secured a monopolistic position that allows them to charge extortionate prices for their products. In

2013, it came to light that it was cheaper for an Australian to buy a plane ticket to the United States and buy Adobe Photoshop CS6 there than it was simply to buy it in Australia.

Intellectual property has been gaining traction in the international treaties we associate with neoliberal globalization. This is no mere recognition of the rise of the knowledge economy but rather a legal lever meant to promote speculative gains—that is, those that have allowed Western countries to maintain their position at the center of a geopolitical context that is increasingly unfavorable to them.

There is an unsettling parallel between the evolution of the copyright industry and that of finance capitalism in recent decades. Historically, the rise of speculative practices has often been linked to the final stages in declining rates of return. In other words, the finance economy comes into play with the greatest force when actual production loses strength as a source of profit. The origins of recent economic deregulation lie in policies developed by the Western economic elite in the mid-1970s to ease their growing difficulties in maintaining the levels of profit they'd enjoyed up to that point. The loss of traditional ways of making money through production makes other business ventures—dangerous, destructive ones like secondary markets and currency speculation—all the more appealing to governments and investors.

Similarly, the digital revolution turned intellectual-property speculation into a profitable business at the precise moment that corporate gains ceased to be the direct result of the production of cultural content. Everyone agrees that digital media constituted a ticking time bomb for the copyright industry. Once users gain access to a master file, it is only a matter of time before it begins to circulate through unofficial channels, whether these are commercial (like the sale of pirated DVDs on the street) or not (as in p2p file sharing). Still, the first digital medium sold on a massive scale—the CD—seems to be the goose that laid the golden egg. It allowed the copyright industry to sell goods far cheaper to produce than vinyl records and cassettes, and sell them at nearly three times the price. Many of the most profitable products were based on catalogs the costs for which had long since been recouped; suddenly, you could get people who already bought that Elvis or Dylan album when it came out to buy it again as a CD at an absurdly higher price.

From that moment on, these speculative practices have spread throughout the economic system, embedding themselves in other activities, from paid cable to the sale of software, via the mobile-phone sector. It is hardly insignificant that entities that manage copyright, like Spain's SGAE, have been caught up in major scandals related to real-estate speculation.[7] The same goes for Italy's SIAE, which was deeply affected by the Lehman Brothers bankruptcy.

Another successful commercial model for gaining profit from digitized intellectual property through financialization is that of distribution platforms like Google, Amazon, eBay, and the Apple's App Store. The secret of these businesses is their size. Through extreme consolidation, these companies are able to make huge sums of money based on the aggregation of infinitesimal profits. In principle, there is nothing illegitimate in this. But the scale of these companies gives them a disproportionate degree of influence that ends up altering cultural supply and demand. They are not simply neutral middlemen; they shape our expectations and those of the producers of culture. This makes for a striking parallel between these practices and currency speculation, in which the enormous amounts of money invested is a crucial element.

This development has dramatically affected the content favored by today's copyright industry. The speculative model for drawing profit from the digital landscape rewards extreme publicity-based consolidation and commoditization and penalizes slower-paced creative activities. Apple has made marketing an art form. The history of alienation reached a milestone when people lined up in front of Apple stores to be the first to buy a product they could easily purchase a few days later and that millions would have within a few months. (In 2011, there were riots at many stores that carried Nike's reissued Air Jordan 11 Retro Concord, but at least that was a limited edition.) Microsoft and Google have developed strategies of hegemonic consolidation that put Coca-Cola and McDonald's to shame and that various organizations above suspicion of communist sympathies—like the European Commission—have repeatedly called into question.

What's so bad about marketing? Advertising affects different products in different ways. There are certain goods and services that could not survive the turboconsumerism of today's societies. Publicists have proven that one can successfully promote commodities that do not, in principle, seem particularly attractive—hybrid cars or even bicycles, instead of sports cars and SUVs—but there are intractable limitations to this, limitations that have to do with the conditions surrounding certain kinds of creation. One clear, though not very glamorous, example of this in the analog world is the recent transformation of the publishing industry. Though it is hard to make generalizations, the work of traditional publishing houses—even those that made huge profits—was fairly artisanal in nature. There was always the occasional bestseller, but publishers dedicated real energy to building a readership for respected authors and genres. What is more, they did not look down on works that sold moderately well over long periods of time, such as essay collections and academic texts.

Today, the publishing industry is fully integrated into the casino economy. CEOs have taken the place once occupied by editors-in-chief. The goal of the big publishing houses, which have undergone a striking process of consolidation, is to come up with a bestseller that produces huge short-term capital gains. This is why they publish dozens of authors and titles they will drop if they don't see immediate results. Marketing plays a fundamental role in this process. Those books that are not likely to have much of an impact in the short term, like poetry collections, are discarded by the industry.

This dynamic affects more than just what books are available for purchase. It has also profoundly transformed what reading is. Until the 1950s or 1960s, most national literary canons consisted primarily of poets and essayists. Today it is the novelists, and not necessarily the most daring among them, who hold this central position. This is not a question of elitism. I am, in fact, an avid reader of science fiction and murder mysteries, and I can easily imagine a world without Artaud or Gadamer being one worth living in. But these commercial strategies create an important feedback loop within the set of practices related to reading and writing in our culture.

It is true that in the case of contemporary popular music, lower production costs—along with newly democratized channels of distribution, communication, and promotion—have produced striking contrasts. Rather than a new model of production, however, the above represents an expansion and renovation of noncommercial, nonprofessional networks of production. Though no one really talks about it, these were dynamics that already existed, up to a point. For example, fans of hardcore punk developed a small but solid network of distributors, listeners, and fanzines at the absolute margin of the industry. Even bands with major international clout, like Fugazi, set contractual limits on how much concert promoters could charge for tickets to their shows.

A good example of a similar cooperative practice was the northern soul scene. Toward the end of the 1960s, young fans of soul music in northern England started getting together to spend their weekends going to clubs where it was played. They were working-class youths who would travel far and wide for these events. After a while, soul's greatest hits started to sound a bit repetitive to them, but they weren't very interested in what the music industry was offering at that moment. Their solution was to comb through the backcatalog of small U.S. record labels that specialized in soul in search of records that had not made it onto the charts. The golden age of northern soul saw the mass importation of singles that had passed unnoticed through the North American market but that English fans valued highly. Northern soul is unique in that it produced practically no music of its own but rather fed off the thousands of records left behind by the industry in its consumerist flight forward.

In the case of both hardcore and northern soul, noncommercial or nonprofit distribution was grounded in very compact communities. Today this sort of circulation is theoretically possible without relying on a local scene. The Internet made it possible to reach individual users spread out around the world. The fact, however, is that the Web has not actually created a virtual community of this sort but has instead fed parasitically off existing offline scenes.

It would be naïve to imagine limitless growth for these non-professional models, even within the cultural sphere. They seem incompatible, for example, with artistic contexts like highbrow music and ethnomusicology. In both cases, production costs are extremely high. The rehearsals required for an orchestra to perform a complex piece can take a long time and need a stable roster of musicians. Ethnomusicology involves extensive research often paid for by public institutions. But this does not mean that highbrow music is condemned to that strange mix of bureaucracy and the star system that characterizes its current model of distribution in Europe. Venezuela's national network of youth orchestras, also known as "El Sistema," proves this point. In the same way, ethnomusicology has been enriched by extraordinary nonacademic and nonprofessional interventions like those of Violeta Parra. It seems reasonable to think, however, that there are situations in which the euphoria for collaborative, nonprofit endeavors would come up against systemic limitations.

Returning to the world of the book, the mediation of cultural gatekeepers plays an important role for which there is no substitute. Listening to a pop song and deciding whether it's worth your time is a relatively quick process. A few seconds are often enough to decide whether we're interested, and so it is feasible that non-specialists could take the place of recording studios, at least for some of their functions. Evaluating novels or essay collections is a much slower and more complex process. If each of us were forced to choose which books were worthwhile from among all the potential offerings of writers who believe their work should be distributed (which is technically possible), our written culture as we know it would disappear. Publishing houses reduce this noise, which is something the Internet is not exactly good at doing.

User comments on the Internet have begun to replace specialized criticism and advertising as essential factors in the construction of literary taste. At first, this seemed like a democratizing turn that would free us from the tyranny of the market and of experts. Reality, however, quickly dashed those hopes:

For some time, writers (or would-be writers) praising their own work under pseudonyms has been an increasingly common practice in online forums, thanks in part to the anonymity of the Internet. . . . At the opposite end of the spectrum, there were users who would anonymously leave scathing reviews of books written by people they clearly felt hostile towards. . . . Both reviews and rankings on sites like Amazon have become parameters on which the success of e-books depends; as such, publishing houses and, above all, self-published authors have developed all manner of strategies to get higher rankings for their books.[8]

The Internet has not made either the book business nor specialist reviews disappear; instead, it has turned amateur reviews into shady business. There are companies that offer reviews on Amazon in exchange for money. For example, GettingBookReviews.com used to offer twenty positive reviews for five hundred dollars. John Locke, the first self-published writer to sell one million e-books, hired the company to get nearly three hundred reviews on different platforms.

Financialization has also affected scientific advances. Here, things are less clear because technoscience remains an important source of production profit. It is true that high-risk investments are introducing more and more bias into research, favoring lines that will be most profitable in the short term. Nonetheless, though the result might not be ideal, it is surely an exaggeration to talk about the phenomenon in the same way as a subprime mortgage or abuse on the part of a monopoly.

This does not mean that the sector is immune to the spirit of the times, and not only in terms of biopiracy. The dotcom bubble fired the starting shot of patent trolling. Patent trolls are businesses that create a portfolio of licenses by purchasing patents from bankrupt companies or patents that have never been used. Their objective is not innovation. They monitor the market in search of businesses to sue on grounds of infringement on the patents they hold. In this way, they make astronomical sums from legal action without investing a dime in research. As such, patent trolling is a parasitic activity similar in form to speculation. Financial firms make direct profit by distorting the function of the secondary market, which they supposedly created in order to make production more dynamic. Patent trolls make profits by distorting laws created to protect scientific innovation. This is no small matter: It is estimated that between 1990 and 2010, patent trolls cost companies engaged in innovation five hundred billion dollars.

Patent trolling has a long history, but right now it is growing at an extraordinary rate. More and more institutional speculators are introducing patent trolling as a natural part of their business models. Major hedge funds buy up licenses on a massive scale in order to sue other companies systematically. To protect themselves, big businesses also buy up huge portfolios of patents, overheating the market. We are witnessing the beginning of a patent speculation bubble. For example, in the summer of 2011 Google bought Motorola's mobile division for an outrageously high sum, in a move reminiscent of the dotcom bubble of the 1990s. The reason: They urgently needed to acquire more than 17,000 patents after losing a bid for Nortel, a bankrupt company that held more than 6,000 patents, to a group of investors that included Microsoft and Apple.

The relationship between the financialization of the economy, the transformation of intellectual property into public property, and the shift in the content commoditized by the copyright industry has not been fully acknowledged by advocates of free culture. It is often suggested that the industry's resistance to new technologies and copyright regulations that are actually more aligned with their potential is the result of institutional laziness. From this point of view, information and communication technologies offer excellent business opportunities within the reach of those content-producing companies that can evolve and adapt to new technological conditions. The copyright industry is an analog dinosaur that refuses to adapt to a new, frictionless context in which size no longer matters.

Comparisons are often drawn to the crisis surrounding musical scores. At the beginning of the twentieth century, the cornerstone of the music industry was the sale of musical scores on behalf of the composers the publishing companies represented. The same laws governed that material as governed books and magazines. With the invention of the phonograph, publishers of these musical scores demanded that the laws remain unchanged, meaning that any artist who wanted to make a recording of a song would have to request explicit permission to do so from the copyright holder. Instead, a law that optimized the social benefits of the new technology was implemented.

> [Anyone] who paid a music publisher two cents would have the right to make one piano roll of any song that publisher published. . . . There was a new technology—the phonograph—that offered the public unprecedented flexibility to listen to music when and where they wanted. There was an old set of rules governing copyright that stated that publishers could control all uses of a musical score they put out, which made it impossible to use this new technology. The answer? A new set of rules surrounding copyright that treated the new technology as a solution, a cause for celebration, rather than a problem that needed to be solved.[9]

Cory Doctorow is both right and wrong here. The decline in the sale of musical scores and the current crisis in the sale of cultural products are entirely different situations. The problem today is not that certain artists are getting rich while others are not but rather that, at least in certain fields, fewer and fewer artists are able to make a living from their craft. The reason for this is that under capitalism, product-related technological innovation—inventing a new router, for example—has a very different economic effect than process-related innovation—such as inventing a way to buy airline tickets without a travel agent. The first tends to increase profits, the second to reduce them. Of course there are still businesses that make money based on the fact that people listen to music, like supplying ADSL or selling headphones, or even speculative practices like the ones described above. The crucial difference is that none of these ways of making money—in contrast to the traditional reproductions industries, whether of musical scores or audio recordings—bears any organic relation to the production of cultural goods.

Many producers of culture have, in recent years, tried to make a profit, but the way in which headphone manufacturers are indifferent to what music their clients play is entirely different from the pragmatism and commoditization of the classic copyright industry. There is an obvious distinction between making a John Ford movie, a Beatles CD, and a book by Tolkien—to name just a few bestsellers—and supplying broadband or selling advertising space.

Even if digital conversion were a viable option for some authors and businesses, it is not clear whether this would be true for all the activities we value. Different institutional contexts affect different kinds of intellectual products in different ways. For example, in Spain the supply of books published is far greater than the demand, with more than a hundred new releases per day. At the same time, bookstores are often permitted to return books to the distributors at a relatively low cost, if they do so within a short time. The result of these combined dynamics is the penalization of works that sell more slowly and the overheating of the literary market. Mystery novels, though often very long, can be dispatched within a few hours. Collections of essays, which tend to be read much more slowly, are pulled from bookstores before it is possible to know what their fate might have been in the medium or long term.

Something similar could happen in the digital sphere. Some artists might choose to adapt by sacrificing material that could have garnered admirers among those who may be unable to show their appreciation for lack of institutional channels through which to do so. This is exactly how certain types of market failure occur, and the similarity is not coincidental. Staunch defenders of commerce assert that pricing is a far more efficient means of coordination than any centralized system. But this is only one side of the story. Pricing transmits information related to scarcity and competition but conceals information related to cooperation, surplus, and decision-making processes.

It may be that the digital sphere as we know it is not the ideal environment for producing and distributing quality cultural products. The Internet may be able to disseminate and provide compensation for science-fiction novels but not for poetic prose; it maybe a suitable platform for smartphone games but not for computing theory. One does not have to be an inveterate pessimist to acknowledge that some of the most brilliant minds of our time are engaged in incredibly puerile activities. According to the computer scientist Jaron Lanier,

> One finds rooms full of MIT PhD engineers not seeking cancer cures or sources of safe drinking water for the underdeveloped world but schemes to send little digital pictures of teddy bears and dragons between adult members of social networks. At the end of the road of the pursuit of technological sophistication appears to lie a playhouse in which humankind regresses to nursery school.[10]

Technological Darwinists avoid passing judgment on content itself, leaving the market to decide what deserves to survive and what doesn't. This strategy is key to understanding the conflicts within the digital sphere today.

COPYLEFT UTOPIANISM

Critics of the copyright industry maintain, quite reasonably, that the digital revolution should be good news. Information and communication technology has the potential to generate tremendous positive change in the dissemination of knowledge and the arts. Though these ideas are often professed with millenniarist fervor, they are hardly controversial. Bill Gates probably believes the same.

These critics seem to believe that the only problems created by digitization arise from the artificial barriers erected by the copyright industry and from a restrictive notion of individual freedoms. As soon as these sources of false scarcity are eliminated, abundance and social harmony will reign supreme. In fact, the problem is different, and very real. The technological options for digital distribution maintain a complex relationship with the varied social contexts of production and consumption defended by opposing political programs.

It is a problem that can be traced back to the very origins of the emancipatory movements that have been trying to outdo utopianism since the nineteenth century. Supporters of socialist alternatives to capitalism insisted that their proposals were within material reach for industrialized societies and were in keeping with their cultural reality. In fact, they presented the project as a fuller realization of the two-part modern revolution in politics and technology. Socialism would provide bourgeois liberty, equality, and fraternity with some real content, while at the same time making more efficient, more rational use of the technological advances developed under capitalism. In other words, socialism presented itself as an alternative that was consistent with the situation before the revolution. But this very idea of continuity implied the need for a process of transformation: a constructive experience through political action that entailed significant practical costs and moral dilemmas. The end of capitalism would not come from some industrialist abracadabra, as the utopian socialists believed.

As such, it is not surprising that contemporary opponents of the copyright industry diverge from the arguments of the traditional Left in this arena. Throughout the twentieth century, the Left demanded that the relationship between creators, middlemen, and the public interest be adjusted to favor the latter. This strategy often involved forging alliances with artists. Not all artists, of course, just those who did not benefit from the commercial system of remuneration and whose cultural practices differed from hegemonic ones.

In terms of how artists are paid, the copyright market is a pyramid scheme. Moreover, its focus on product differentiation through brand management means that it tends naturally toward homogenization. A caricature of this is that the industry gives us a choice between Cristina Aguilera and Britney Spears, Lady Gaga and Kesha, Coldplay and the Killers, but not between Alban Berg and Ghanaian highlife music. This is why criticism of the copyright industry has typically adopted simultaneous positions regarding how artists are paid, the democratization of both access to channels of production *and* distribution, and also of the content deemed worthy of circulation.

It would be absurd to argue that these positions have always— or even often—been consistent; far more so that they have been healthy. The Left has a long and motley history of artistic conservatism, avant-gardism, populism, and cultural elitism. It should be said, however, that it has traditionally engaged at least three areas that are completely beyond the contemporary criticism of the copyright industry: the free circulation of content, the social conditions for real access to information (not just making this hypothetically possible), and the question of how artists are to subsist. The position of the Left traditionally involved emphasizing public or communal communication networks and promoting noncommercial systems of remuneration for artists and pedagogical structures as alternatives to marketing.

The approach of contemporary critics of the copyright industry is quite different and in some cases actually the opposite. It centers on two related points, the first of which is ethical and the second of which has to do with the way production is organized in society. What the two points have in common is that both are individualist and procedural. This is not necessarily a bad thing; in fact, many on the Left have enthusiastically welcomed the change.

At first glance, it seemed that copyleft had achieved the objectives of oppositional cultural movements and cleared up several irresolvable ideological dilemmas. By not addressing content itself and instead offering a noncommercial alternative with no centralized coordination, it seemed to free itself from some of the most frustrating arguments in the leftist tradition: What is a revolutionary cultural practice? Is bureaucracy really better than the market? The search for an alternative to the market economy often led to proposals reactionary in their substance and ineffective in their implementation. Copyleft seems to preserve the best of both the market and its noncommercial alternatives. It encourages individual creativity, allows for cooperation, and limits both bureaucratic control and commoditization. What could possibly go wrong?

The fact that copyleft originated in software development is not insignificant. One specific conflict is paradigmatic of the free-culture movement: the creation of a completely open operating system known as the GNU project.

The story is well known. In 1983, the programmer Richard Stallman announced his intention to develop an environment compatible with Unix—a reliable and widely used operating system—that would not only be developed in a cooperative spirit but that would be guaranteed never to be privatized in the future. For this reason, the project needed a special license that not only asserted that any user could run, copy, modify, and distribute it but also one that prevented all future restrictions of those rights. The idea came to be known as copyleft.

Copyleft is based on four interrelated freedoms: the freedom to run the program, the freedom to study how it works and change it, the freedom to distribute copies with or without the intent to make a profit, and the freedom to distribute copies of the modified version to third parties. These four freedoms are "viral" in nature: Anyone can exercise them over any product licensed in this way, but the same license must apply to all products derived from a copyleft-licensed product. Copyleft licensing is not limited to software; it can be applied to any kind of intellectual property. If, for example, someone wanted to sell a new and improved print-based version of Wikipedia, they could do so without asking anyone's permission, but that version would have to permit the same freedom of use as the original.

Copyleft was fundamentally shaped by its origins in the world of computing, which makes its adoption beyond that sphere somewhat surprising. One could imagine it might be desirable to be able to, for example, modify a mathematics textbook to suit the needs of certain students or expand on certain points. What is less clear to many, on the other hand, is how freedom is served by making changes to nonutilitarian texts like poems or philosophy essays. These issues, however, should not be blown out of proportion. Modification is not uncommon in some areas of art, such as works written for the theater. Plays are often adapted for creative or logistic reasons. A theater company in a men's prison, for example, might eliminate female roles from a play in order to put it on.

In the case of free software, modification is a fundamental and inalienable element. The implications of this are not only legal but also technological, as it means open access to the programs' source code. Richard Stallman himself summed up his GNU project as follows: "The principal goal of GNU is to be free software. Even if GNU had no technical advantage over Unix, it would have a social advantage, allowing users to cooperate, and an ethical advantage, respecting the user's freedom."[11]

The primary characteristic of copyleft is, in effect, a commitment to breaking down the barriers that limit the flow of information, in a broad sense. As a declaration, it is highly regulatory and sets strict procedural guidelines. These procedures do not include an a priori criterion for identifying a positive outcome. From their perspective, the result is good if it respects the rules. One example of a nonprocedural criterion can be seen in the remarks of the Chilean dictator Augusto Pinochet, who announced that he would accept the results of a democratic election as long as no one from the Left were elected to office. For supporters of copyleft, limiting access to information is toxic, no matter what the end. Because of this, true copyleft regulations include the freedom to distribute the licensed material for any purpose, including financial gain.

Copyleft challenges the copyright industry only insofar as the latter is built on an information monopoly. In fact, copyleft openly opposes certain anticommercial practices typical of the Left. Licenses that allow reproduction as long as there is no commercial use involved are not copyleft. This has generated major conflicts in the world of free culture. Most musicians, writers, and filmmakers critical of the copyright industry employ licensing that allows free reproduction only for noncommercial purposes, for example through the ubiquitous Creative Commons Attribution–NonCommercial–ShareAlike license. Software developers, on the other hand, often allow their programs to be used for any purpose.

To be fair, the computer scientists who created copyleft insisted from the very beginning on the difference between the freedom the system allows and whether it is free of charge. "Free as in 'free speech,' not as in 'free beer' " is a slogan repeated often in the software world to address the ambiguity of the word. In this context, the way the information was produced and how it will be used are not considered relevant at the time of licensing. This is the source of the tension between defenders of copyleft and the worlds of music and publishing: The social conditions governing the payment of programmers—many of whom are salaried employees or could be if they so wished—are entirely different from those of musicians, for example, who are often self-employed and receive a percentage of each work sold.

Procedural standards make this a difficult subject to broach, and I believe this fact has limited the spread of free licensing. There are cases, such as those of independent musicians, in which granting others the freedom to reproduce works for commercial purposes is not a viable option. However, the opposite also occurs. Certain forms of artist compensation should *reduce* conflict over free licensing. In the case of most public orchestras, for example, musicians impose major restrictions on their performances: Even the use of a recording just a few minutes long needs to be authorized by a committee. Yet these are public employees who receive a reasonable salary, work under more than acceptable conditions, and whose continued employment is practically guaranteed. Their performances might reasonably be expected to be in the public domain, in the strict sense of the term. Then there are other subsidized works, for example, films, pieces of visual art, and doctoral dissertations supported by grants. Does it make sense to assign a restrictive license to a fully subsidized film or to charge the public-television system to transmit it?

In general, there is a clear tension between the efficiency with which copyleft blocks the privatization of the commons and its supporters' resistance to considering the social conditions of production and the uses of those freedoms. Historically, the process of privatizing the resources of the commons under capitalism has played an important and multidimensional role as a mechanism of appropriation by dispossession or as an essential factor in the emergence of a modern workforce. Copyleft takes direct aim at these processes.

But the expropriation of the commons is not the only path to monopoly; in fact, it is not even the main path. Marketing, economies of scale, privileged information, consumer myopia, and the complicity between politicians and big business: Copyleft supporters have decided not to address these issues, or rather, they have decided to treat them as an uninteresting byproduct of restrictive laws. They speak to us of a world in which small-scale entrepreneurs and artist cooperatives can challenge big business, in which creativity and hard work are appropriately remunerated. From this perspective, the only obstacle is the legal chain-link fence constructed by monopolists.

In this sense, copyleft bears an unsettling resemblance to recent neoliberal deregulatory strategies. According to copyleft, the problem is not the information market, much less the labor market; the problem is barriers to the circulation and use of information. Commercial relations are one possible channel for the transmission of information and, as such, are not harmful in and of themselves. It is the licenses that are harmful. Once this idea has been accepted, criticizing monopolies and fees is only a step away from denouncing all barriers to commoditization. Not all supporters of copyleft are fans of the market, but many of them see this as a personal choice and not necessarily related to the freedom of information. In the end, it turned out to be "'free' as in 'free market.'"

Much criticism has been levied against viewing the market as an efficient system of distribution. One very reasonable complaint is that pricing tends to homogenize diverse goods and services that require different approaches to distribution. As a procedural strategy, copyleft reproduces this leveling of distinct products or spheres: the patent for a vaccine, political information, a videogame, a word processor, a song . . .

◉ ◉ ◉

The Left was not always so categorical regarding the free circulation of information. I do not believe that this means it was not especially committed to freedom of expression. Quite the opposite. It simply did not focus on procedural matters. In fact, this is not even solely a characteristic of the political Left. During an electoral debate in Germany toward the end of the 1960s, for example, the Social Democrat Willy Brandt accused the Christian Democratic Union candidate Kurt Georg Kiesinger of planning to authorize private television stations in West Germany. An agitated Kiesinger denied the accusation, arguing that private stations would mean the end of German democracy.[12]

Indeed, until not long ago it was widely thought in Europe that private television stations were a significant threat to freedom of expression, even when they coexisted with public ones. It was thought that television had a huge influence on public opinion but did not promote reflection the way that other media, like newspapers, did. For this reason, it was deemed necessary to protect the public sphere from the impact of private television stations, which were thought to be less committed to the truth and more subject to commercial pressures.

According to contemporary media ideology, this is a fairly paternalistic position. Perhaps it is. And while it is not as though public television behaves that much better, or even all that differently, it is also true that these reservations about commercially based media pluralism are relatively prudent. When faced with regulations that could prove harmful, focusing on procedural questions without considering the outcome of the process is both naïve and irresponsible. This was the error of financial speculation: The question is not whether secondary markets are useful in theory but rather what risks they pose and what consequences they might have in a world inhabited by human beings. Similarly, it would not make much sense to evaluate the monopolistic privatization of the airwaves solely in terms of whether it is lawful without taking into account the social processes that would likely follow and that would be practically irreversible once set in motion.

In general, it is simply untrue that free access to information creates greater critical consciousness. Morozov explains that one fascinating discovery made by the leaders of the former DDR was that those cities that had access to Western television were more satisfied with the communist regime than those where there was no reception of West German channels.[13] *Dallas* did not help undermine the dictatorship; it fortified it. Many authoritarian regimes today are remarkably tolerant of access to Western entertainment. The Communist Party of China has realized that Lady Gaga is not an enemy but rather an ally.

One could, similarly, question the commercial use of knowledge generated in the public sphere. Attaching a copyleft license to a vaccine developed in a public institution could have very different consequences depending on the economic context: A cartel of laboratories could make the vaccine exclusively to sell it at a high price in countries without a public-health system able to produce and distribute it. A publishing house could, likewise, systematically take translations published on the Internet under a copyleft license and publish them on paper in less technologically developed countries where they already dominate the book market.

For the Left, the monopolistic concentration of information is incompatible with democracy. Most advocates of free knowledge are against having information fall, de facto, into just a few hands; however, few proposals have been made to reverse the situation by establishing consequences along the lines of antitrust laws. The reason is that this would go against a legal position based on a strictly procedural understanding of freedom of expression.

The second defining characteristic of copyleft is cooperation. By eliminating restrictions on intellectual property, copyleft incentivizes the creation of productive spaces that spontaneously come together through encounters in collaborative networks. There are both intellectual and social dimensions to this. In neither case are value judgments made regarding the content of the material produced, the way the material is used, and the conditions of its distribution.

The intellectual dimension has to do with the idea that, according to the Californian creed, the Internet is a space in which fragments of information converge, forming what is often described as a "hive mind." The two most widely cited examples are Wikipedia and the collaborative, noncommercial development of free software. What is almost always forgotten is that both projects are idiosyncratic and cannot be universalized; their particular features are not shared by the majority of technological, cultural, political, productive, and scientific practices.

The development of major software programs is always collaborative—this characteristic is not limited to free software. Software development can and should be divided up. A whole mythology has been built up around independent programmers working all night, alone in their garages. In reality, breaking a large project down into sets of problems to be solved collectively in a kind of assembly line is not only an option—it is a necessity. It is impossible to imagine that, for example, Windows 95 could have come about in any other way. The unusual thing about programming is that, unlike many traditional production processes, it does not require strict temporal or spatial continuity. Also, compared with other kinds of informal collaboration, programming involves technical criteria that—at least to a point—make it possible to settle disputes or at least reaffirm the authority of certain individuals.

Making a conventional movie, for example, has virtually none of these characteristics. It requires that many people be in a given place at a given time, under specific circumstances (technical requirements, weather conditions, and so on). Many of these people—the actors, at least—are not substitutable, except for certain minor roles, and there is no strict need to adhere to any specific technical procedures. Certain methodologies may be deeply engrained, but if someone decides to bypass them, they are not going to end up with a compilation error. Pasolini shot some of his greatest works without having the slightest idea about cinematic conventions. Other cultural practices share some, but not all, of these qualities. Another example from the world of the screen are television screenplays, which tend to be written collaboratively, though there are hierarchies and time pressures at play. An animated film, for its part, is potentially much easier to break down into chunks.

The second example that often comes up is Wikipedia, a cooperative publishing project in which a vast number of anonymous contributors participate under equal conditions and with the aid of a series of technical tools and editing guidelines. The mechanics of Wikipedia is very simple: Any user can modify or create an article as they see fit. Other users can in turn reject, correct, or expand that work, and so on. You need only search for an entry, and if it turns out not to exist or is in some way substandard—incomplete, sloppy, poorly written—you can create it or modify it yourself.

There are many unusual aspects to creating an encyclopedia, however. Again, any large-scale encyclopedia—not just Wikipedia—

is bound to be a collaborative effort. Generally speaking, encyclopedias—unlike essays or opinion pieces—aim toward neutrality. Unlike monographs, they do not present original arguments but are instead secondary texts based on original work. Encyclopedias collect knowledge accumulated and agreed upon by a community of scientists and experts. In this sense, the suggestion that Wikipedia is an example of the success of collaborative work on the Web hinges on circular logic. Setting aside the number of collaborators, the same could be said of Diderot and D'Alembert's *L'encyclopédie*.

In fact, the distinctive thing about Wikipedia is not so much the fact that it is collaborative but rather that it is an encyclopedia compiled by nonexperts: The debates surrounding its content are horizontal, and a student has just as much right to edit an article as a tenured professor. Additionally, Wikipedia has far fewer cognitive barriers to entry than free software does, at least in theory. For Internet-centrists, Wikipedia proves that online we each contribute the bits of expert knowledge we possess to the aggregate. A teenager in Spain might not know anything about the water regime of the Guadiana River, but she can correct a spelling mistake in the name of one of the towns mentioned in that topic's entry because she used to spend summers there with her family. Supposedly, the moral of the story is that intelligence is collective and made up of tiny individual units. With the right tools, pieces of microknowledge can accumulate, giving rise to new levels of understanding.

The magic of information networks is that they allow these cognitive fragments to collect in an orderly fashion without centralized coordination. In fact, it is often said that they do so more effectively than if a central node were controlling them. This argument is highly debatable, at least in the case of Wikipedia. A few very optimistic evaluations of the quality of Wikipedia articles are based on the voices of specialists. It can be assumed that the users who wrote the entries "Wavelet" and "Sine wave" are experts. In this sense, Wikipedia could be said to be a parasite of conventionally organized academic institutions. Knowledge may be a collective endeavor, but it is highly debatable that Wikipedia provides any definitive evidence in this regard.

This is also true of another frequently cited example of collaborative intellectual activity. Crowdsourcing is an open invitation that brings a large group of people together to perform tasks that traditionally would be carried out by one individual or an organized collective. The results have been rather mixed—failures abound, but there have also been some real success stories. Foldit, for example, is a computer game created by the University of Washington. It is a bioinformatics simulation designed to predict the structure and folds of proteins based on their amino-acid sequences. The program was created so that people with no specialized knowledge of biochemistry could help find the natural forms of proteins. Those who want to participate play a videogame in which they move geometric shapes around in a competition to find the most efficient configuration for the protein: The less energy your structure requires, the more points you get. In 2011, using this method, they were able to determine the three-dimensional structure of the enzyme of a retrovirus similar to HIV, after 200,000 players generated 18,000 unique designs.

Foldit, however, is not a knowledge-aggregation system like Wikipedia. It is instead a case study in consensual social manipulation. The cooperation extends only as far as the decision to play the game. It is as though a local power company were to hook dynamos up to all the exercise bikes in the area and then use the energy they generated. One could not exactly call that "collaborative energy production." Foldit is interesting because it brings to light the way in which many tasks are more multifaceted than they seem and require unexpected abilities (for chemistry majors, there is no special course in folding or in playing videogames). This need not present any ethical dilemmas, but we should not confuse the program for something it is not. What it resembles is cracking a code by testing out a huge number of combinations using a powerful computer or distributed system. This can sometimes be an effective method, but it is not the same as deciphering it through reverse engineering. In fact, Foldit was founded on the—very true—idea that when it comes to three-dimensional spatial reasoning, humans are more powerful calculating machines than computers.

Regardless of whether the idea of a hive mind is accurate, it is gaining influence as a metaphor. There is a fairly clear parallel between intellectual cooperation and, again, the sui generis forms of collaboration generated by the market. Traditional forms of collaboration develop either through face-to-face interaction or within the confines of regulated institutions and organizations. The idea that cooperation could be a byproduct—intentional or not— of other interests is far more outlandish. It is tied to the universalization of the market in the modern age, which spread the doctrine that social coordination arises spontaneously from individual, self-interested action, with no need for institutional intervention. There is a pronounced symmetry between the idea that knowledge is granular and the view that pricing is the ideal means for optimizing resource allocation.

According to the Californian creed, once the proper channels of transmission are established, noncentralized intellectual aggregation occurs naturally: Knowledge is the result of the spontaneous coordination of fragments of information. Collective intelligence looks a lot like an intellectual free-market system. People may believe they are contributing to collective knowledge or not, and they may be willing to do so or not. It doesn't matter. The hive mind is a byproduct of interaction.

There are at least two serious problems with this view. The first is shared by all those who idealize the market as an organizing force. In fact, there has never been—and, quite feasibly, will never be—anything remotely like a free market that is both enduring and widespread. Neoclassical economists argue that this is because we haven't made enough of an effort. For them, the failures of the free market are attributable to insufficient zeal in purging public interventions and monopolies.

It is hard to think of another economic system that has maintained itself with such obstinacy despite its startling instability and devastating externalities. The truth is, talk of deregulation has always been superficial in that the market depends on interventions from the state or other institutions both to eliminate the natural tendency of human beings to coordinate in noncompetitive ways and to limit the social damage wrought by commoditization. The main contribution of the ideology of pricing is to make these interventions seem irrelevant, reframing them as exceptional cases rather than as the historical norm of capitalism.

The hive mind is also like this. When you scratch the surface of the examples of so-called collective intelligence, you immediately see that digital cognitive processes are quite similar to traditional ones. In fact, it could be argued that these ventures found success because, regardless of all the talk about their reticular structures, they actually look a lot like conventional collaborative endeavors. They found success in spite of digital technology rather than because of it.

Wikipedia, for example, is not just an encyclopedia. It is also a community of users much smaller than the image of its being made up of countless grains would suggest. Although millions of people use Wikipedia and thousands contribute to the site from time to time, the number of people who spend time improving it is actually very limited. Some of them are administrators, users with editorial privileges who make some of the most important and polemical decisions on Wikipedia. In this sense, it would not be an overstatement to assert that the administrators are the core of Wikipedia's sociological community. As a result, Wikipedia's workflow ends up strongly resembling that of a traditional, noncollaborative encyclopedia, given that the administrators often behave the way editors do in the world of the conventional book. This has actually helped improve the quality of the encyclopedia, but it is also a source of conflict insofar as there is a contradiction between this pragmatic reality and the ideological principle of popular, granular, anonymous collaboration espoused by Wikipedia and defended by the administrators themselves.

The second problem has to do with motivation. Defenders of pricing claim that the quest for personal gain has the unintentional side effect of generating a greater collective well-being than would have resulted from coordination or altruism. Collective well-being, then, does not necessarily originate from an individual quest for virtue. Worldly ethics are compatible with social behavior. Giving in to selfish impulses may seem morally objectionable, but it is not necessarily injurious to society as a whole. Can the same be said of the motivations behind cooperative networks? Why do people collaborate online?

Copyleft promotes cooperation not only as a vehicle for the aggregation of knowledge but also in a more conventionally social sense. It paves the way for people to undertake projects together and motivates them to do so. This is surely the aspect of copyleft that has most contributed to its popularity among the Left. The absence of legal and material barriers fosters this kind of cooperation. If I have access to the source code of a program or the text of an encyclopedia entry, I can manipulate and eventually improve them. I don't need to begin every project at square one, or view it as mutually exclusive with other projects, because my contribution forms part of a chain of interventions. On occasions, this input might be coordinated formally through an organization (for example, a group of mathematics professors writing a textbook together), but the interesting thing is that this is not strictly necessary. Though my contributions may be highly sporadic, and though I may not have the slightest contact with the other contributors, I can still be part of a cooperative community. Information and communication technology is essential to this cooperation. A math textbook written by several authors can be written analogically, but it is much more difficult to do so if there is infrequent personal contact between them.

This matter is more important than it might seem. One of the key criticisms socialists levied against the market system was that, within it, economic freedom stopped at a corporation's front door. That is, though unlike under feudalism, a salaried employee is free to accept a job or not, those who do accept have to submit themselves to companies' internal rules and regulations, which are generally extremely vertical and hierarchical. Businesses are archipelagos of authoritarianism surrounded by a superficially egalitarian legal context, and we spend half of our waking adult lives inside them.

Bringing democracy inside company walls is, however, quite costly. The creation of an effective labor community, like a cooperative, requires a delicate combination of personal commitment and institutional architecture. Not all decisions need to be subject to democratic deliberation—those dealing with technical or urgent matters, for example, do not. Moreover, intense social ties do not necessarily make for an ideal work environment. There are advantages like loyalty and commitment, but there are also disadvantages, like the difficulty of negotiating.

One of the reasons copyleft has been so successful among the Left is that it seems to vastly reduce the cost of horizontal collaboration. The same arguments that are used to assert that information technology renders intellectual collaboration automatic by turning the aggregation of knowledge into the unintentional result of other processes are also employed to claim that they lower the cost of social cooperation. The lack of personal ties allows individuals with different levels of engagement with the project to come and go from collaborative projects as they please. Cooperation can take place on a massive scale because it is not limited geographically and because the ripple effect of social media allows each participant to be linked in to many others. In other words, a significant portion of the Left has taken up one of the dogmas of cyberideology: the inherent ability of information technology to facilitate social interaction.

In reality, the idea that technology can help strengthen and broaden connections between individuals is fairly outlandish. The history of the past three or four centuries—which is, in part, the history of technological change on a seismic scale—is marked by a progressive weakening of social relations as humanity once knew them. Social scientists are nearly unanimous in linking modernization with the destruction of traditional community bonds. Until the eighteenth century, most men and women had a pretty clear sense of what their life would be like, where they would live, at what age they would marry, and what their profession would be. Industrialization, commoditization, and the growth of cities— like democratization and the Enlightenment—tended to dissolve the symbolic bedrock that shaped individual lives and collective decisions in the past. This has allowed many to free themselves from having their lives laid out in advance and has created a wealth of opportunities, but it has also generated insecurity and disorientation.

It is generally understood that technological advances are a stimulus, if not a direct cause, of the fragmentation of experience and social ties. Adam Smith admired the way a needle factory divided labor into miniscule tasks, but at no point did he think that this approach could benefit sociality or personal development. The extreme division of labor characteristic of technologically advanced economies makes it hard for us to get a full sense of the tasks we perform at our workplaces and how these relate to other aspects of our lives.

According to the sociologist Richard Sennett, this dynamic has become more pronounced in recent decades. Widespread commoditization has led to an erosion of personality, of subjectivity. It is not only labor processes that have been fragmented: Occupations and professions themselves no longer bring a sense of coherence to our lives. In general, nothing does. According to many sociologists, there has been a radical transformation in personal identity, that is, in the way we understand ourselves. We no longer see ourselves as a coherent continuum that is connected to a more or less permanent physical and social context but rather as an incoherent chain of heterogeneous experiences, fleeting emotional connections, unrelated jobs, impermanent homes, and conflicting values.

Information technology is paradoxical in this sense. According to today's ideology, its effects are the exact opposite of those of traditional technology. Of course, this is not to say that it should motivate us to return to traditional subjectivity and social relations; rather, it would be the seed of a new generation of social bonds able to mend the fragile social ties that characterize modern life.

It makes sense that many political activists are tempted by this idea, which seems to respond to the socialist longing for a form of community compatible with the modern demand for personal freedom and autonomy. Socialism sought a model of fraternity that preserved individual liberty, and copyleft seems to be the epitome of that ideal: individuals involved in a wide range of collaborative activities without forming personal dependencies. Through copyleft, we would finally arrive at a virtuous cycle consisting of individual freedom and creativity, strong communal foundations, and economic development.

Nonetheless, these aspirations to free knowledge may well re-
semble the liberal idea of sociality. Or one of its versions, at least.
The historical legitimization of capitalism was not only grounded
in Hobbes's anthropological pessimism. Not all defenders of commerce viewed society as a zero-sum game. On the contrary, the
market was also understood as a solution to the brutal oppression
and conflict typical of societies where political foul play abounded.
This is precisely the idea behind the phrase "sweet commerce," which
was coined by Montesquieu in the eighteenth century to denote the
way businesses could form social relations that, while superficial,
were friendly and calm. He saw the market as an alternative to the
great religious and political passions that had turned Europe into
one vast battlefield in the early years of the modern age.

Many Enlightenment thinkers were suspicious of social relations and believed that commerce mitigated the natural tendency
of cultural and political differences to degenerate into open conflict. They hardly believed that commerce was the ideal backdrop
for the realization of human virtue, but they did view it as a lesser
evil, a better alternative to political or religious war. Montesquieu
says explicitly in *The Spirit of Laws*:

> Commerce is a cure for the most destructive prejudices; for it is
> almost a general rule, that wherever we find agreeable manners,
> there commerce flourishes; and that wherever there is commerce,
> there we meet with agreeable manners. . . . Happy is it for men
> that they are in a situation, in which, though their passions
> prompt them to be wicked, it is, however, their interest to be
> humane and virtuous.[14]

These thinkers had a clear memory of the bloodbath Europe became as a result of political and religious conflicts. Some of them
believed that commerce could foster amicable ties that, while perhaps less virtuous than the political relations of Athens or Rome,
were also less aggressive. The bet placed on commerce was, deep
down, the result of historical decline. Constructive politics was
possible for contemporaries of Pericles or Solon, not for the Europeans of the eighteenth century. In the age of Louis XV the quest

for political greatness tended to lead to disaster. Better to opt for the social ties characteristic of business deals—they were shallow and catered to the lowest common denominator, but at least they were calm and cordial. At its core, Montesquieu's proposal was to foster political stability by lowering the bar of social expectations.

The European Union has similar origins. The founders of the European Coal and Steel Community (ECSC), the precursor to the EU, explicitly sought to create shared commercial interests in parts of Western Europe as a means of preventing military conflict in the region. A vast array of political and cultural efforts had not managed to keep the historical enmity between France and Germany from dragging the planet into two world wars. Commerce would work that miracle.

In the age of casino capitalism, it is hard to maintain this faith in the social power of the market. The Internet, however, presents itself as an opportune surrogate. No one would claim that having a Facebook friend or a follower on Twitter is the same as a real friendship, but in the context of generalized social fragility perhaps it is the closest we can come. For those who champion the present, this might even be a step forward: the chance to reinvent ourselves and make the most of our creative abilities, free of anthropological burdens. According to a widely held opinion, the glue that binds our society together is produced in a space online in which autonomous individuals with no relation other than their shared interests come together. The key is that the social bonds forged through information technology can coexist with the postmodern fragmentation of subjectivity. What is more, the bonds depend on that technology.

Anonymity and immediacy allow us to collaborate, share, and participate in a community if we want, when we want, and with the identity of our choice. On the Internet, a series of discontinuous subjectivities coexist that have no past or future beyond their preferences in the moment. Information technology breaks empirical personalities down into a series of compartmentalized identities and, above all, provides a technical mechanism for reconstructing social activity by means of participatory instruments. Traditional social relations are replaced by diffuse and discontinuous—but augmented, technologically enhanced—connections. We might not have extended families, close friends, or professional careers anymore, but information travels in far broader circles. Participation in the technological sphere is the vector that consolidated the extreme plasticity of our individual identities. Facebook users unite . . . in being Facebook users.

The secret to this cybersociality is, as in Montesquieu's commercial cordiality, to lower our expectations. In reality, Web 2.0 tools have not solved the problem of fragile social ties in modernity or the postmodern fragmentation of personality; they have made them more inscrutable through the dissemination of technological social prostheses, the same way that the widespread use of psychopharmaceuticals did not do away with the experience of industrial alienation but instead simply made it less conflictive. Information technology has created an impoverished social reality rather than an enhanced one. For the first time, mass culture is not just a metaphor. The Internet has not improved sociality in a postcommunal context; it has simply reduced what we expect of social ties. Nor has it contributed to our collective intelligence—it has simply lowered the standard for what we consider to be an intelligent remark (140 characters is a pretty modest measure).

This is why, as Jaron Lanier explains, the rise of connectivity, social networks, cloud computing, and sharing has led to a profoundly negative exaltation of the dynamics of mass culture, more similar to José Ortega y Gasset's reactionary nightmares than to communitarianism. In a completely deinstitutionalized digital context, the simulacra of sociality—Facebook "friends"—and collaboration—the "likes" announced on digital media homepages—emerge as if by magic from individual, voluntary convergence in online spaces. Lanier points to how the hegemonic model of technology is diminishing our view of human personality:

> The attribution of intelligence to machines, crowds of fragments, or other nerd deities obscures more than it illuminates. When people are told that a computer is intelligent, they become prone to changing themselves in order to make the computer appear to work better, instead of demanding that the computer be changed to become more useful.[15]

Can a conversation in a chat room be considered a social bond in the same way as a family or peer group can? Isn't it a bit like comparing political freedom with the freedom to choose which products to buy? Above all, why would the idea of lowering the bar for sociality work any better in the case of technology than it did with commerce?

COOPERATION 2.0

There is a strange paradox to open-knowledge movements. On one hand, they overestimate technology's potential. Technological advances are never independent of the social context in which they occur, and a radio may be a much more effective means of communication than a computer in certain situations. On the other hand, these movements can be strangely atavistic in many of their claims. It is fascinating how little is said in cyberutopian circles about processes that affect millions of people, like unemployment, lack of political representation, gender inequality, and the financial crisis—especially when compared to the popularity of minor events much further removed in time and place.

Experts have drawn analogies between digital-rights management (DRM), access-control technologies that limit the use of digital devices, and the legal process of enclosure, which facilitated the expropriation of common land in England between the seventeenth and nineteenth centuries. They also see similarities between the generosity found online and the potlatch, traditional feasts held among the Native Americans of the northwestern United States until the start of the twentieth century. They ask us to think of the Internet like a bazaar, the secular institution of commercial exchange with origins in Persia.

This is not a minor point. It reveals how most scholars of technology radically turn their backs on the problems of contemporary society, as though the Internet allowed us to reconnect with the polite and comprehensible world of traditional societies by bracketing off the violent and irresolvable contradictions that characterize industrial ones. This tendency has also infected cyberactivism. Perhaps this is why the only alternative to commoditization that has emerged from its ranks has been the recovery of the commons, a historical relic whose main virtue is that it does not oblige us to specify the institutional model within which it should come about.

The commons is particularly amenable as an intellectual backdrop because it belongs to societies with limited technological development that have disappeared or are on the verge of disappearing. This is the most convenient way of avoiding uncomfortable questions like: Which is better, a cooperative system with a professional structure that operates within a commercial setting, like the federation of worker cooperatives that is the Mondragon Corporation, or an anarchist alternative that breaks radically with hegemonic systems, like ecovillages? Is centralized planning the alternative to the market? Can there be competition without profit motive?

Why would someone altruistically decide to spend their time programming, translating, subtitling, writing, and sharing music and films? Some cases are easy to explain, such as those of paid content providers on file-sharing sites and authors with no other way of publishing their work. In many other cases, however, this work is carried out anonymously and without profit motive. This is not a traditional social relationship, but it would be reductive to consider it in terms of the superficial ties typical of consumerism.

Most of us cooperate frequently only with others in our immediate circle: our children, our parents, our friends. This type of interaction is based on face-to-face relationships. There are certain idiosyncrasies to it, like the noninterchangeable nature of those who take part in these relationships. If a sibling or friend dies, we can't simply find a replacement in some database.

In modern societies, we also see large-scale models of impersonal collaboration; the two most important of these are the labor market and state bureaucracy. Both require an elaborate institutional framework with all sorts of rules, means of coercion, knowledge, and physical infrastructure. Salaries, for example, are in fact an extremely complex arrangement designed to organize forms of coordination based primarily but not exclusively on self-interest rather than networks of interdependence.

In most cases, digital collaboration is not based on traditional interpersonal relationships, formal organizations, or self-interest. How does information technology manage to foster cooperation within the most tenuous or nonexistent institutions and without personal relationships? The most common answer is that they facilitate altruism. The market is good at enabling cooperation grounded in self-interest, though it is a poor vehicle for generosity and concern for the well-being of others. Family and peer networks often include some element of altruism, though not always and not necessarily. We cannot generalize this principle, however, and treat everyone like a sibling. Online cooperation seems to offer the best of both worlds: universality and altruism.

Information technology produces a kind of market for altruism, a trafficking of talents. On the one hand, online interactions do not depend on self-interest as they do in the market. As the story goes, there once was a contractor who wanted to get rid of a huge pile of sand left over from a project, so he put up a sign that said: "Free sand." No one wanted any. The next day, he put up a new sign: "SALE: 100 lbs of sand for one cent," and the sand disappeared in no time. In the market—and, by extension, in other traditional spaces of commerce—one cannot be altruistic. Not because it would be bad; it is simply that altruistic motivations are incompatible with the regulatory framework of the market. It's a bit like one scene in the Argentine comic strip *Mafalda*, in which Mafalda's friend Felipe is given a chess set as a gift. Felipe tells her that he doesn't play chess "as well as Najdorf" and adds, "He must have much better aim than me."

There is virtually no commercial space in which I can, for example, give a book away free of charge. No bookstore in the world would agree to manage my books for free, and with good reason. Sometimes, however, simply breaking with the vocabulary of commerce is enough to allow altruism to emerge. In one case, a pensioners' group reached out to the local bar association to inquire whether any of its members would be willing to offer a discount to elderly individuals with limited finances. No one responded. The pensioners' association then asked if the lawyers would be willing to offer their services for free to those in need. Many agreed.[16]

Noncommercial interpersonal relationships depend on stability, and it is often expected that we demonstrate some degree of concern, at least in certain situations, for others. Our online interactions are sporadic and involve very little personal investment. The Internet, however—unlike the market—does offer a space for altruism.

This is possible because, from an academic perspective, there is not much difference between altruism and egoism. Rational-choice theorists tend to study self-interested behavior because self-interest is more straightforward than altruism. Altruism can be reduced to self-interest, but it doesn't work the other way around. This operation might distort the deeper meaning of altruistic behavior, but it is methodologically sound. In terms of its decision structure, altruism is an individual preference like any other. Altruism consists of putting the interests of another before one's own; egoism is simply the opposite. Economists believe that there is no reason to view these preferences as two substantially different types of behavior, in the same way we see no reason to view liking sports cars as a behavior in strict opposition to liking SUVs. Both altruism and egoism can be explained as the result of a hedonistic calculation, that is, the quantity of satisfaction we derive from behaving in a certain way.

This is consistent with a depressing discovery in cognitive psychology: We are much more compassionate when it comes to tragedies that affect us subjectively than those we consider objectively more serious. It is not true that access to more information leads to greater solidarity and altruism; in reality, it nearly always leads to less. The things that increase the likelihood that we will be concerned about others are those that generate empathy: an image of a sick child, for example, rather than a statistic about the millions of children who die of malaria each year. This would seem to suggest that, if sociality is not limited to empathetic face-to-face relationships, then it is not based on altruism (an individual's concern for others).

This point is worth considering in greater detail. In our daily lives, we act according to two very different types of behaviors: instrumental and normative. According to the concept of instrumental rationality, you are behaving rationally if you choose (what you believe to be) the best means at your disposal to get what (you think) you want. This is the kind of behavior that is expected of us in the market. From this point of view, the desire itself is irrelevant when it comes to classifying a behavior as rational: it could be a desire for the greater good, or to kill and dissect the last living member of a species on the verge of extinction. At the other end of the spectrum, normative behavior is based on a system of shared rules whose origins are not entirely clear and that cannot be reduced to instrumental rationality. This is the kind of behavior that governs our family life and friendship circles.

Jon Elster offers an example that may prove illuminating. Let's say that John Smith is willing to pay some kid ten dollars to wash his car, but not a penny more. If the kid were to ask for eleven dollars, John would spend half an hour washing the car himself. Now let's say that John's neighbor offers him twenty dollars to clean her car. One can easily imagine John indignantly refusing the offer. The mysterious forces that have made him abandon the valuation of eleven dollars that he just placed on a half-hour of his time are in fact social norms.

Because the distinction between regulations and rationality is so elemental, we tend to understand them as a clear-cut dichotomy. In fact, it makes much more sense to think of them as the two extremes of a continuum. Social norms with clear utilitarian dimensions often include an instrumental component. For example, because family relationships are a key element in traditional economies, dowries tend to be very important when it comes to establishing a matrimonial bond. This was not because the ancients were cold-hearted, selfish, and unable to form emotional relationships with their spouses but rather because they did not make a categorical distinction between the domestic and the economic. The paradoxical consequence of this was that the economy, as we understand it today, was less central to people's lives and that family relationships were more protected. At the other extreme, we tend to think that adherence to ethical norms is worth less if there is some instrumental factor behind it. We do not trust a witness who is rewarded for his testimony as much as one who testifies in spite of the risk it implies.

Orthodox economists try to explain everything in terms of instrumental behavior because it is simpler; ultimately, it follows a truly basic logic. All it takes to behave according to strict instrumental rationality—choosing what is considered the best means to a given end—is a simple computer program. Based on elementary logic, one can arrive at mathematically complex calculations of questionable utility (what we call academic economics). Norms, however, can vary according to context and interpretation; worse still, we don't have the slightest idea how they come into being.

This is why the prisoner's dilemma is so interesting. It is something of a fable of the limit case of a group whose members behaved exclusively according to the criteria of individualistic instrumentality. The dilemma is this: If all the members of a group behave according to rational self-interest, they will be worse off than if they do not. On the other hand, from an instrumentalist perspective, none of them has a reason *not* to act out of rational self-interest. If no one else cooperates, it makes no sense for *you* to do so; if everyone else does, it makes more sense to take advantage of them. If no one pays taxes, it is absurd for me to do so because my sacrifice will mean nothing; if everyone does, it is still absurd for me to pay because I can just mooch off the rest.

These vicious circles are rarely seen in the real world because groups tend to establish regulations that govern collective endeavors—like paying taxes—and institutions that enforce cooperation and penalize freeloaders—like tax collectors. The problem is that it has been shown that these kinds of norms and institutions cannot arise from instrumentalist calculations; they imply a radical shift in perspective. The moral of the story is that, beyond any doubt, the most efficient social interaction is in some cases irrational from an instrumentalist perspective. The so-called solutions to the prisoner's dilemma try to limit the number of noninstrumentalist norms necessary for cooperation to occur as much as possible, at least in theory.

Generally speaking, the prisoner's dilemma clearly demonstrates the distinction between social norms and instrumental rationality and shows that a narrow conception of individualism is too limiting. If real life worked the way the prisoner's dilemma assumes it does, society as we know it would not exist. Sociality is tied to norms and institutions that cannot be reduced to individual beliefs and desires.

As I said before, instrumentalist behavior is individualistic but not necessarily egoistic. It matters little whether, in my practical reasoning, I place my personal preferences ahead of the preferences of others. Formally speaking, the decision structure is identical. As such, individual altruistic behavior is just as subject to the prisoner's dilemma as self-interested behavior. Imagine, for example, that a couple rob a bank. They are arrested and held without a way to communicate with each other. The police only have circumstantial evidence against them, and if neither confesses, they can only be sentenced to one year in prison. If one confesses and the other does not, the one who confesses will be sentenced to ten years in prison, and the other will go free. If both confess, the prosecutor will be lenient and recommend a five-year sentence for each. The two are madly in love, and each wants the other to go free, with no thought for themselves. In that situation, they will both confess, and both will be sentenced to five years. In any scenario, confessing seems to yield the best result for the other. If both do so, however, the outcome is worse for the other than it would have been if they had managed to work together to save themselves.

Adhering to social norms, for its part, can be utterly egoistic, insincere, or malicious. One follows the rules for whatever reason one sees fit, and it doesn't matter; the important thing is to follow them. The true opposite of egoism is not altruism but rather compromise. The idea of compromise suggests the unique way we adhere to norms that cannot be reduced to instrumental rationality. This is not always, or even often, a question of major moral choices. At one end of the spectrum, we follow the rules simply to follow the rules. For example, we accept proper dining etiquette without really considering its purpose. We do this because it is what is done: Norms bind us to certain behaviors. Whether one follows these rules happily or not, the essential thing is our commitment to these obligations, not the pleasure they bring us or the beliefs we hold about them. As Tony Soprano said to his teenage son, who was going through a Nietzsche phase and refused to go to church with his family: "God may be dead, but you're going to kiss his ass, anyway." Jon Elster offers a more refined historical example:

> Under "mature" communism . . . everybody knew that nobody believed in the tenets of the official ideology, and yet everybody was compelled to talk and behave as if they did. . . . The reason why the leaders forced people to make absurd statements in public was not to make them believe in what they said, but to induce a state of complicity and guilt that undermined their morality and ability to resist. Indeed, they were so hollowed out as individuals that, as a woman from the former East Germany said, "she could not just suddenly 'speak openly' or 'say what she thought.' She did not even really know precisely what she thought."[17]

It is often the case that obligation strongly influences altruistic behavior. This is why the two things are often conflated. But if these normative obligations don't depend on altruism, on what do they depend? Basically, on interpersonal relationships and institutions. Both of these limit the desire, the possibility, the opportunity, and the benefit of going against the system, whether in the form of selfish freeloading or of taking a moral stand against an injustice.

The fact is that when we adhere to a norm we are not acting like irrational automatons—we can consider various alternatives, such as following the rules partially or not at all. Moreover, norms tend not to be unambiguous or straightforward but rather highly dependent on context. We can fool ourselves into thinking that we are following the rules or that our infraction is minor or justified. This is why so many systems of social norms include procedures and mechanisms to ensure they are being observed, along the lines of "kill your firstborn with a flint axe during the summer solstice or the assembly of Just Men will stone you both to death." The combination of regulations, procedures, and supervision is roughly what we call an institution, a codified way of doing something that should not be confused with an organization or a community (that is, a collective agent).

The relationship between norms and institutions is fairly clear, whereas the relationship between norms and communities is much less so. In general, there are norms that affect empirical communities, in which personal relationships tend to be important, and norms that affect abstract communities. In the latter case, the community may end up being no more than the sum of its norms. Classical sociology speaks of primary and secondary organizations. This is actually a problematic distinction: The most reasonable approach would be, again, to view the thing as a continuum. At one end would be those practices that bear no perceptible relationship to the community, like having good table manners or adhering to the social norms that dictate our behavior while standing in line. At the other would be organizations with strong emotional elements, like family relationships. Our commitment to the first set is weaker than to the second, in the sense that we often comply with these norms simply because there is no real opportunity or no good reason not to.

Most people adhere to several different sets of norms. One typical trait of simple societies, however, is that these sets can be organized into a comprehensible hierarchy—an imaginary one or one based on self-delusion, perhaps, but one that is at least coherent. There are connections between the different strata of rules and obligations.

In modern times, on the other hand, the standard state of affairs has been confusion. Sociologists describe our societies as individualized, but this is not entirely true. Most of us are pathologically committed to organizations—businesses, above all. Historically, few religions have boasted devotees in numbers as vast as ours, the salaried workers. But these circles of loyalty are fraught with complications. We break our backs for organizations to which we should only be connected through calculated self-interest, and we neglect our inner circles, despite singing their praises with a degree of sentimentality that would have seemed absurdly saccharine to anyone born before the nineteenth century. In this sense, describing our societies as "complex" isn't entirely accurate—we should instead probably say that they are "chaotic."

The solution we have found to deal with this bleak reality is bureaucracy, in the sense established by Max Weber. We delegate the creation of explicit interpersonal codes that regulate certain aspects of societal collaboration to a group of experts. For regulation of this sort to function, some organizations must have coercive power. The difference between these forms of bureaucratic cooperation and traditional rules is not so much the regulations themselves but rather the level of personal involvement—low in the former and high in the latter. This is what distinguishes a modern army of salaried soldiers or mass recruitment from, say, the system of rules that turned free Athenians into hoplites, citizen-soldiers who collectively provided military security. This distribution of commitment is not necessarily a bad thing. Do urbanites really want to be as involved with the workings of our city's water supply as a farmer is with his traditional irrigation system?

Orthodox economics assumes that instrumental rationality is the foundation of human behavior. Nonetheless, a strange finding in experimental psychology suggests that one of the few groups that systematically follow this pattern are economists—and professors and students of economics. The tremendous influence of this historically outlandish and morally toxic understanding of human behavior is linked to the disproportionate power we have given to the few people who consider it significant. This is something we in the West have some experience with. After all, the sexual mores we followed for so long were established by celibate members of the clergy.

Online collaboration demonstrates, in case anyone ever doubted it, that we are not systematically selfish. Many people choose to share things and dedicate their time to others with little incentive or social pressure to do so. On the Web, our concern for our fellow man can be infinitely sporadic and is not tied to any stable regulatory structure. At first glance, this seems to pose no problem. Indeed, it seems to resolve a dilemma that is characteristic of complex societies.

For many, capitalism not only has major social and material shortcomings; it also presents a broader problem in terms of the kind of motivation it demands: fear, self-interest, competition. The market allows for the coordination of certain human endeavors without interpersonal relations of dependence. The results are, at best, a mixed bag. It is true that the market has helped eradicate the vestiges of certain oppressive traditions and, generally speaking, helped promote a unique brand of independence and individual freedom. The price we've had to pay has been the destruction of certain characteristics we consider important in people, such as concern for other people. The digital sphere, for its part, is character-ized by an individualism and an anonymity not unlike those of the market but that do not require us to turn our backs on others. The Internet allows us to be monads, but that doesn't mean we are condemned to rational egoism.

There is, nonetheless, one crucial limitation. Cooperation in the digital sphere depends on altruism, understood as a personal choice, rather than on obligation, understood as a social norm. Merely participating in the digital sphere does not automatically generate cooperation. It is something I can choose to do or not, and I must justify my choice. Let me illustrate this difference by way of an anecdote.

I used to get together with a group of professors for a mid-morn-ing coffee. Without ever explicitly agreeing on it, we'd adopted the practice of taking turns to pay the group's check. There was no es-tablished order; every day, someone would offer to pay, and the re-sult was more or less a regular rotation. It wasn't much money, and no one really minded if it didn't work out to be exactly even. Still, it was hard not to notice that one professor never offered to pay. Time went by, and the situation grew increasingly uncomfortable, but no one stepped up to confront her about her behavior. After all, there had never been a formal agreement about taking turns to pay the check. Suddenly one day, after another colleague had offered to pay, the stingy professor said, "Wait, wait. You always pay." "Finally," I thought, "she's figured out that she has to pay every now and then." But to my surprise—and, to a certain extent, admiration—she added firmly, "Today, we should go Dutch."

The stingy professor refused to acknowledge that she was par-ticipating in a cooperative system of norms based on reciprocity (a common arrangement in many traditional societies). To her, it was a matter of altruism, of concern for the others, and, as such, a personal choice that she could make as she saw fit. She didn't want to pay for anyone else; her concern for the rest of us was limited to not wanting us to spend money on her.

Likewise, no one passes judgment on the amount I choose to donate to a cause: I become altruistic with the first cent I give. In contrast, there tends to be a minimum threshold for organized cooperation. If I remove a few blades of grass from around my front door, I am not participating in the shared labor of cleaning the streets of my town. In fact, any suggestion to that effect might be seen as a provocation. There was a panhandler who used to beg for money outside a supermarket near my house. When someone would try to give him small change, he would pull back his hand and exclaim, indignantly, "I don't take copper!"

The idea that there is a relatively high minimum standard for cooperation poses a challenge to the theory of basic rationality, one that is well known. The experiment that best illustrates this point is Ultimatum, a game played between two people who do not know each other and will never see each other again. The first player proposes how to divide a given sum of money—one hundred dollars, let's say—between him- or herself and the second player. If the second player rejects the proposal, neither gets a penny. If the second player accepts the proposal, the money is divided accordingly. Economic rationality tells us that A will offer the lowest possible sum, one cent or the like, and that B will accept the offer, since one cent is better than nothing (a kind of "take the money and run" attitude). Yet it has been proven that in most cases A will offer a substantial sum, often nearly half, and that B will reject all offers that fall much below that mark. The experiment has been repeated in different cultural contexts, with very similar results.[18] On the Internet, however, just like in the market, the absence of thresholds is perfectly acceptable. Crowdfunding is based precisely on this "take the money and run" logic.

Similarly, the need to justify cooperative behavior is socially uncommon. Many systems of norms include altruistic behaviors. The most important thing about these rules, though, is that they don't ask us to come up with reasons for following them. In fact, beyond a certain point the search for explanations ends up destroying these rule systems, as theologians well know. If I ask myself earnestly and systematically why I don't take the Lord's name in vain, I'll end up jotting down enough notes to formulate a true skeptic's response. If I ask myself earnestly whether I should pay taxes, it's very likely that I'll end up in jail for tax evasion.

At a certain point we follow the rules, and that's all there is to it. As the philosopher John Searle explained, I can't just walk into a bar, have a beer, and say to the bartender, "Listen, while I was having that beer I analyzed the situation carefully and, to be completely honest, I've found that I really don't feel like paying you." Stepping into a bar binds us to a system of norms that include paying for what we consume, whether or not we feel like doing so. Likewise, and fortunately for newborns, we don't need to crave changing our children's diapers in order to actually do so. To commit oneself to the care of a child means forgetting about desires and preferences and behaving more or less appropriately on a regular basis.

There is no system of rules online that acts on me in this way. Initiatives in digital collaboration have been very imaginative when it comes to defining intelligent and effective rules of operation. Analog communities could learn much about institutional innovation from free software, Wikipedia, and P2P, but no empirical digital communities elicit a commitment, in the strict sense of the term. This is why every now and then we see a message from Jimmy Wales urging us—with good reason—to donate to Wikipedia. It sounds very civilized, but the fact is that if caring for others depended entirely on personal motivation, sociality would be impossible.

Most successful labor cooperatives in the analog world have strong community ties. The Mondragon Corporation is one of the biggest cooperative projects in the world and one of the ten largest business groups in Spain, bringing together 280 businesses and boasting a strong international presence. Even so, it is very much geographically grounded near the Basque town of Mondragón and has an integrated network of professional development and research centers, and even a cooperative university, there.

The case of the Mondragon Corporation suggests that a stable system of cooperation is more like an ecosystem than a calculation of costs and benefits. For better or for worse (and often for worse), it has to do with personal and social identities, with what defines who we are and what we aspire to be. This barely exists on the Internet because it is easy for me to cut off social interaction. If I systematically sabotage the conversations in some forum or other, the worst thing that can happen to me is that I get kicked out. Platforms like eBay and Digg offer tools that allow their users to evaluate one another and establish reputations. Destructive behavior might ruin my digital identity in those forums, and I might have to give up my handle. But it is difficult to compare these online consequences with the sanctions of our peers in the analog world and the way these affect how we see ourselves. The only times this cost is higher is when some kind of massive backlash from my virtual actions affects my analog self. The so-called Streisand effect is not simply anecdotal. It suggests that the Internet can generate social effects similar to those of an analog community and not all that different from an angry mob.

The interesting thing is not so much that there is no powerful system of norms on the Internet but rather that there is good reason to believe that there never could be one, in an organized sense. Likewise, there is not and could never be anything resembling bureaucracy online, though not because it is actually impossible. Some government might give it a try, surely at an astronomical cost, but the result would be something other than the Web as we know it, to which decentralization is fundamental.

The price we pay for the Internet's combination of independence and cooperation is that it cannot be a bastion of self-governance in any real sense. We participate in anonymous altruism as long as it doesn't require too much commitment. The production of free content online is often parasitic, in the sense that it depends on external sources of income and free time. As the joke goes, the best way to make money as a developer of free software is by being a waiter. No one is willing to risk their life, in the broad sense, for an anonymous and potentially fickle crowd that doesn't even respect the basic tenets of anthropological reciprocity.

Many people who collaborate online do not see themselves this way, and for good reason. They believe themselves to be honestly committed to the dissemination of knowledge and well-being. Their cooperative activities are, without question, an important part of their lives. Someone once told me that on one of Richard Stallman's first trips to Spain, he was given several CDs by local bands. He politely refused the gift, explaining that he did not want to possess any material goods that bore restrictive licenses.

The problem is not ethical integrity, a sense of personal responsibility or coherence, but rather the existence of a system of norms that governs cooperative activity in a stable and effective way without condemning it to the vagaries of individual choice. I think that many people sense this limitation, deep down, which is why the expression "the commons" appears so often in the jargon of cyberactivism.

The commons are those resources and services that, in many traditional societies, are produced, managed, and used by the whole community. It could be pastures or crops, water, fisheries, hunting, tasks related to the maintenance of roads, harvesting, making pottery, or caring for dependent members of the group. This system has been known by innumerable names throughout history: the commons, *tequio* and *minka* in the Americas, *andecha* and *auzolan* in Spain. Contemporary sociology tends to call them common-pool resources (CPRs). Cyberactivists claim that there is at least a formal similarity between these secular forms of cooperation and composing an article on Wikipedia, programming free software, or altruistically subtitling movies or television shows. Does this make sense? Why are the commons conceptually important in a context so unlike their original one?

This discussion brings us back to Garrett Hardin's well-known article "The Tragedy of the Commons," which explains how the administration of shared resources faces a significant dilemma. Essentially, if several individuals—acting rationally and motivated by self-interest—use a finite shared resource independently, they will end up depleting or destroying it, despite that it is in none of their best interests to do so. This is another version of the prisoner's dilemma. The two orthodox solutions that are generally put forward for this tragedy of the commons are either privatization or bureau-cratization. The logic behind privatizing the resource is that each owner will ensure the preservation of the part that belongs to them and will guard against any mooching by their co-owners. In bu-reaucratization, an external body manages the resource, oversee-ing how it is allotted and punishing any infractions.

A typical response to this argument—one that is not very sound—is to accuse Hardin of applying circular logic. The problem of the commons only arises if those involved behave out of modern rational egoism, but that is not the way members of the traditional societies in which common property existed tended to act. In fact, Hardin had a much more dynamic understanding of the problem than he is generally credited with. It is true that he is a little short on historical details (he was a Malthusian zoologist), but it is not hard to reimagine his arguments in more sociologically precise terms: Can the commons survive in a complex society—that is, in a deregulated context?

As such, and despite what is often said, the economist Elinor Ostrom did not refute Hardin. Instead, she posed another, equally interesting question: How were the commons able to survive in tra-ditional societies? Members of Neolithic societies were neither ethical heroes nor idiots blindly following collectivist ideals like sheep. They were able to identify as well as we can what was in their best interest, or in that of their families and the community, and were often tempted not to adhere to social norms. In fact, the mys-tery is actually how the tragedy of the commons was not a more frequent occurrence. In other words, the surprising thing is that incredibly stable communal systems administered shared resources for centuries without relying on coercive external agencies.

Through her ambitious empirical investigation, Ostrom estab-lished the institutional conditions under which stable and effec-tive communitarian agreements regarding shared resources are likely to emerge—a sophisticated organizational framework that traditional communities developed over time:

> "Institutions" can be defined as the sets of working rules that are used to determine who is eligible to make decisions in some

arena, what actions are allowed or constrained, what aggregation rules will be used, what procedures must be followed, what information must or must not be provided, and what payoffs will be assigned to individuals dependent on their actions. . . . One should not talk about a "rule" unless most people whose strategies are affected by it know of its existence and expect others to monitor behavior and to sanction nonconformance. In other words, working rules are common knowledge and are monitored and enforced.[19]

Ostrom also identifies several "design principles" shared by long-standing CPR institutions.[20] Basically, the individuals and families affected by the system must be clearly identified; the rules regarding use and distribution must be consistent with the local context; the participants must be able to modify these arrangements by collective choice; there must be ways to monitor and sanction behavior and resolve conflict; finally, the participants must have the right to organize, and the existence of nested collective entities must be possible.

Many CPR systems with these characteristics achieved equal or better results than systems created through individual competition or by a government agency. They are the result of a slow process of evolution, not accidents or the product of trial and error. That is, this is not a case of blind submission to collectivism or unconditional altruism. In fact, the examples Ostrom cites tend to involve long-term deliberations that encompass a wide variety of motivations.

The limitation of Ostrom's argument is that she primarily studies traditional communities, many of which found elegant and effective solutions to their organizational issues. Is it possible to draw an analogy between these examples and present-day digital collaboration? In a word: no. The Internet as we know it presents practically none of the conditions observed by Ostrom nor ever could. To this point, I would assert:

1. There are few clearly defined systems in place in the digital sphere for managing shared resources. One rarely knows with any

certainty which people or collectives have the right to a given resource and who is in charge of its distribution. Wikipedia, for example, is an open environment with a decidedly heterogeneous array of collaborators: regular encyclopedists, occasional contributors, and belligerent individuals known as trolls, who participate only in their fields of interest (topics related to their political beliefs, for example). This is a source of real problems, which are being countered with strategies like not allowing anonymous users to create new profiles. There are a few extremely closed communities—forums for pedophiles or hackers, let's say—in which trust is a key element. It is significant that these spaces are associated with criminal behaviors. They create a form of negative obligation—analogous to Isaiah Berlin's negative liberty—which does not emerge from shared responsibility but rather from a negative-sum game: If I lose, you lose.

2. There is a meaningful connection in CPR systems between local conditions and the rules that govern the distribution and use of resources. In a community that shares water resources, water is distributed differently in years of drought and in years of heavy rainfall. Digital media are, by nature, expansive and unaffected by context; they tend more toward compartmentalization. As I have already pointed out, one common cause for debate among programmers and writers is that a strict copyleft license, typical of free software, is not generally as problematic for practical texts like dictionaries or manuals as it is for creative works. The free distribution of works online might be advantageous for artists who have other sources of income, like giving live performances, but catastrophic for those who do not have that option, like film actors. This compartmentalization also prevents most individuals affected by the system's rules from being able to modify them, another characteristic of stable CPRs. The absence of community ties renders collective decision making quite costly in a distributed context. For this reason, many cooperative projects begin as an initiative of a small group, sometimes just one person, who establishes the rules that those who join later will follow. This is also the reason there are so many iconic figures in the sphere of digital cooperatives—Lawrence Lessig and Linus Torvalds, for example—whose influence

transcends their intellectual merit and more closely resembles a personality cult.

3. In CPR systems, appropriators who violate the rules are sanctioned by other members or by specialized functionaries. There are also mechanisms in place to resolve conflict quickly. On the Internet, monitoring and sanctions are extremely costly and inefficient because of its unclear rules and huge scale. Supervisory mechanisms do exist; some are social (like the ratings assigned to news articles), and others are hierarchical (the roles played by Wikipedia editors and forum monitors). These perform as badly as one would expect, and mutual accusations of trolling and censorship—with an angry mob mentality thrown in for good measure—are practically synonymous with social networks.

4. The most complex CPR systems are typically organized into multiple levels of nested entities. The very idea of a distributed network runs contrary to this. There are a few minor nods to the right to organize, but they are fragile and ineffectual. There are, for example, the Wikimedia Foundation and the Free Software Foundation, but their relationship to the projects to which they belong is largely prescriptive.

This set of limitations extends beyond the digital sphere. Many hold an economy of shared resources up as an alternative to neoliberal capitalism. They seem to believe that one can be committed to the commons in general without dealing with the concrete norms that regulate goods and services within that system. This is a mistake. What Ostrom has shown is that using a shared resource is synonymous with following the rules that govern how the resource is managed, just as playing chess implies following the rules of chess. These systems might include different levels of specialization and involvement but not simply a nod to some vague notion of solidarity or a dedication to public property in general. CPR systems are different from both private management and state administration.

Anyone who believes that an economy of shared resources could be compatible with complex societies must also maintain that there are rules governing the distribution and supervision of shared resources that are compatible with high levels of anonymity and the fragility of empirical social relations. Modern societies multiply not only the opportunities and reasons for freeloading but also the complexity of the problems to be resolved. The limitations to large-scale participation in the management of numerous important organizations are cognitive as well as institutional, from the oncology ward in a hospital to the supply of potable water in a big city.

The technical impossibility of taking part in the decision-making core of a CPR system could be a meaningful limitation. We tend to see those goods and services to which we contribute, and the function of which we understand, as being more meaningful. The more marginal our involvement in the process, the harder it is for us to identify with it. This is why well-intentioned attempts to affix participatory appendages to bureaucratic and technically complex processes tend to fail. In the context of urban redevelopment, public budgetary meetings and interviews with residents demand significant time and energy; nonetheless, users remain at the periphery of decisions regarding the actual use, management, and distribution of the goods and services in question.

The moral of the story is that governing the commons is inextricably tied to a traditional understanding of communal life. Rich, prolonged relationships within the community are essential to the success of systems of norms in which the temptation to cheat would be strong if interactions were anonymous and sporadic. Ostrom herself points this out, indicating the limitations of rational-choice theory in understanding CPRs:

> [These models] are far less useful for characterizing the behavior of appropriators in the smaller-scale CPRs. . . . In such situations, individuals repeatedly communicate and interact with one another in a localized physical setting. Thus, it is possible that they can learn whom to trust, what effects their actions will have on each other and on the CPR, and how to organize themselves to gain benefits and avoid harm. When individuals have lived in such situations for a substantial time and have developed shared norms and patterns of reciprocity, they possess social capital

with which they can build institutional arrangements for resolving CPR dilemmas.[21]

One of the keys to the success of a CPR is what Ostrom calls the "incremental self-transformations" that occur in the process of creating the institutions that govern them. The idea is that this process sets off a feedback loop of learning that enriches that institution over time. In contrast, though information technology does much to facilitate communication and dissemination, its institutional characteristics (intermittence, lack of a dedicated space for discussion) mean that cooperative endeavors face a number of complications.

Cyberfetishists believe that the rules of the game change online. They believe that information and communication technologies create a unique kind of sociality based on the intersection of one-off individual actions. Cooperation becomes the common factor in a purely communicative space made up of individuals united only by their shared interests: computer programming, legal questions, hobbies, the search for sexual relationships, artistic creation, the collective writing of encyclopedia articles, and so on. It is not a community based on personal bonds or a shared life project.

In a way, it resembles the bourgeois fantasy of a compartmentalized social contact that leaves the private sphere untouched, the old desire to see the public labor of economics, politics, and culture take place in watertight containers that require no commitment beyond participation in the activity itself. It is also highly consistent with the image of sociality presented by rational-choice theory. For economists, cooperation that does not result from calculated self-interest or an individual penchant for altruism is problematic, even in its most trivial manifestations. Even cooperation with oneself—something most of us take for granted, except in cases of serious mental illness—can pose issues. One well-known example of this is the paradox of the smoker. Given that each cigarette plays an infinitesimal role in the development of possible future illness, the smoker never has a rational motivation not to light up, as the damage caused by each individual cigarette is outweighed by the benefit of the pleasure it provides. Nonetheless, the sum of these acts causes an aggregate damage—a terminal illness—that exceeds the aggregate benefits. Hence the paradox.

The reason behind this shortcoming of rational-choice theory is that it takes our day-to-day empirical identity—our "I"—as a collective. As though the present version of me the smoker were different from the me trying to kick the habit or the me stricken with emphysema who regrets all those years I smoked, even though it is the same person experiencing these states throughout his life. According to this, the proper perspective on one's life would focus on the present moment. What is technically the "I" of rational-choice theory is an atemporal blank that must be constantly filled in anew to avoid internal contradictions. Obviously, real people are not like this. Our obligations to institutions and social norms shape our behavior despite our personal preferences, and that is the basis of our actions in society. This is why economists have such trouble explaining the emergence of institutions like CPR systems, which should technically be irrational but which prove to be quite effective.

On the Internet, however, sociality does not seem to violate the principles of rational-action theory. Computers are a mold that makes people behave like fragmentary individuals. In cyberspace, collaborative projects are based on technological procedures that are apparently indifferent to empirical personal identities. As I observed earlier, this anonymity and immediacy allow us to collaborate, share, and form part of a digital community when we want, if we want, and with the personality of our choosing. Online collaboration seems to be the product of a random assortment of perfect rational decision makers with no past or future beyond their immediate preferences. Information and communication technology creates an ideological veil that makes this possible. It breaks empirical personalities down into a series of compartmentalized identities and, above all, provides a technological means of reconstructing social interaction through participatory mechanisms.

In this sense, the Internet performs a function similar to that of the labor market: It is a practical mechanism designed to liberate collaborative activity—intellectual labor in one case, workforce labor in the other—from the institutional conditions in which it originally developed. A procedure by which a type of bond that in all earlier societies was based on relationships of mutual dependence is transformed into a formal transaction. Information and communication technology creates the fiction of a new kind of community, a new social order made up of fragments of the self, of infinitesimal parts of an identity, in the same way Wikipedia is made of the infinitesimal bits of erudition each user is able to offer.

In reality, online cooperation is as similar to a political community as a large corporation is to an extended family. The Internet is the postpolitical utopia par excellence, based on the fantasy that we have left the major conflicts of the twentieth century behind us. Postmodernists imagine that these symbolic and cultural changes are moving us away from crass liberal individualism, in which the basest forms of self-interest provided the motor for social change. They also believe we have moved past the bid for a welfare state that solves certain problems but drowns creativity in a sea of drab bureaucracy. They imagine a world full of entrepreneurs who guard their individuality while being creative and socially conscious. A world in which knowledge is the core value of a competitive but fair and nonmaterialistic economy and in which the new economic leaders prefer surfing to yachts, home-baked cookies to caviar, hybrids to sports cars, and fair-trade coffee to Dom Perignon.

Cyberfetishists, then, are not just wrong—they've provided a false solution to a real problem. The problem of the commons in a complex society is just a stylized version of the fundamental ethical dilemma of the Left. We want to be free individuals, and at the same time, we want to be part of a network of solidarity and meaningful, not just bureaucratic, commitment. We want an efficient economy that allows us to choose between different professions and incentivizes talent in a way that benefits us all. But we don't want a labor market that forces us to compete and produces inequality.

The problem is that each of us wants to forge our own social ties based on choice rather than obligation, but we want everyone else to form a robust network of solidarity that will protect us and guarantee that this cooperation will be sustained rather than sporadic. It's like what happens when we go on vacation. We travel to places that would be incredible if it weren't for all those people packed into them—people who decided, like we did, to visit an incredible place. In other words, cooperation on the Internet brings us back, rather violently, to the origins of emancipatory movements. The real question is whether we can apply some of what we learned over a century of attempts at social transformation or whether we have to go back to square one.

2

AFTER CAPITALISM

EMANCIPATION AND INTERDEPENDENCE

CYBERUTOPIANISM UPDATES AN idea that is common in modern revolutionary movements: that of improving on the paternalism typical of traditional communities and of the emergence of a form of
social relations based on solidarity while being respectful of free
personal development. Any politically engaged critique of the
fantasies some have about the Internet demands that we also examine the way in which the Left has presented this fundamental
issue. Digital triumphalism is a bad answer to a good question . . .
like many recent ideas about postcapitalist society.

Cyberfetishism is so appealing because it conceives of the present as the result of a fruitful, nonviolent break with the past. From
this perspective, we are the lucky heirs to technological shifts with
important social and perhaps political side effects. The truth is, the
burden we hope to rid ourselves of is a heavy one. Technoutopianism
is not so much a smokescreen as it is a balm of delusion to soothe
the pain of an unbearable historical legacy, in which reality appears
violently excessive. Reason, in this case, doesn't exactly thunder—
to borrow a phrase from "The Internationale"—instead, it is the
inoffensive, banal elevator music coming out of our iPods.

It is hard to say whether our recent past has been an unhappier
time than previous eras. I imagine it would be more or less the same

to die of starvation or exposure in a thirteenth-century Central European village as it would in a German or Russian concentration camp, and there probably isn't much of a difference between having boiling oil poured over you while storming a medieval wall and having napalm dropped on you in the Vietnamese jungle.

On the other hand, the way in which the great challenges and catastrophes of our time generate action rather than just words is historically unique. This action is the result of a social order that is beyond our control but that could be modified, given the right conditions. At least, so it seems to us. As Slavoj Žižek puts it, with his strong Lacanian inflection:

> In contrast to the nineteenth century of utopian or "scientific" projects and ideals, plans for the future, the twentieth century aimed at delivering the thing itself—at directly realizing the longed-for New Order. The ultimate and defining moment of the twentieth century was the direct experience of the Real as opposed to everyday social reality—the Real in its extreme violence as the price to be paid for peeling off the deceptive layers of reality.[1]

The rhetoric of immateriality, the abundance of the digital sphere, reticular social relations, and post-Fordism try to hide the fact that everything today is more or less as it was before two world wars, the fossil-fuel crisis, decolonization, standoffs between blocs, the nuclear-arms race, and neoliberalism. Not in the sense that people have more or fewer problems today—again, this would be hard to evaluate—but rather that problems of the past continue to haunt us, though we pretend not to see them, not so much like in *Scrooged* as in *The Sixth Sense*. The message we don't want to hear is that our hopes for a cyberutopia never saw the light of day.

◉ ◉ ◉

Just over a hundred years ago, the West began to feel the combined effects of a set of destructive social dynamics that stretched back to the heroic years of capitalism, effects that non-European countries had already suffered. Systemic overproduction, colonial expansion, political systems in crisis—the establishment of capitalism created social, economic, and political tensions on a vast scale from 1914 to 1991, a period sometimes referred to as "the short twentieth century," which then spread throughout the world. We are dealing with the same puzzle today, but it is as though we've painted the pieces with cheerful colors and given them high-tech designs. The problem is the same, but it ends up being more confusing and harder to solve.

The fact is, the twentieth century was one of the bloodiest times known to man, at least in quantitative terms. It has been said that 154 wars took place between 1990 and 1993 and that these took more than one hundred million lives, 80 percent of which were civilians. The Italian novelist Erri de Luca—who says he hates to hear someone disparage the twentieth century—often quotes a poem by Osip Mandelstam that captures the secular zeitgeist well: "My age, my beast, who will ever / Look into your eyes / And with his own blood glue together / The spines of two centuries?"

Testimonials from the time reveal just how much influence these events had on people's perception. John Berger recalls that for some time the fear of a nuclear holocaust was so urgent and pervasive that it had repercussions on decisions as personal as one's artistic calling:

> I gave up painting not because I thought I had no talent, but because painting pictures in the early '50s seemed a not direct enough way to try to stop the world's annihilation by nuclear war. The printed word was a little more effective. Today it is hard to make people realise how little time it seemed to us was left to prevent this ultimate disaster.[2]

Toward the end of the 1980s, Martin Amis—a postmodern writer known more for his irony than his political commitment—wrote *Einstein's Monsters*, a collection that is striking in today's context for the fearful, anxious tone with which it approaches the threat of nuclear war, a problem we have frivolously come to ignore.

Political opposition is, likewise, a collective experience inextricably linked to this historical moment. The radical Left spent most of the twentieth century immersed in a practical dilemma that was both completely predictable and unavoidable. The philosopher Gerald Cohen offers this illuminating anecdote:

> In August 1964, I spent two weeks in Czechoslovakia, in Prague, in what was then the home of my father's sister, Jennie Freed, and her husband, Norman. They were there because Norman was at the time an editor of *World Marxist Review* . . .
>
> One evening, I raised a question about the relationship between justice, and indeed moral principles more generally, and communist political practice. The question elicited a sardonic response from Uncle Norman. "Don't talk to me about morality," he said, with some contempt, "I'm not interested in morals." The tone and context of his words gave them this force: "Morality is ideological eyewash: it has nothing to do with the struggle between capitalism and socialism."
>
> In response to Norman's "Don't talk to me about morals," I said: "But Uncle Norman, you're a life-long communist. Surely your political activity reflects a strong moral commitment?"
>
> "It's nothing to do with morals," he replied, his voice now rising in volume. "I'm fighting for my class!"[3]

In *The Age of Extremes*, Eric Hobsbawm talks about people who were even more committed than Uncle Norman. Olga Benario, for one, who was the daughter of a successful lawyer from Munich and who joined Germany's Young Communist League in 1923, when she was fifteen years old. Olga made a reputation for herself in clashes with Nazi militias in the streets until she and her partner, Otto Braun, were arrested. Benario was released and participated in the raid of Moabit prison to free Braun. They later managed to escape to the Soviet Union. After that, Benario traveled to Brazil, where she met Luis Carlos Prestes, the leader of an uprising that spanned the most remote regions of the country. The insurrection failed, and the Brazilian government handed Benario over to Nazi Germany, where she died in a concentration camp. Otto Braun, for his part, ended up in China, where he was the only Westerner to participate in Mao's Long March.

In the 1970s, the same Erri de Luca was in charge of the security force of the Italian left-wing organization Lotta Continua. He describes the violence of Italy's "Years of Lead" as an objective reality, with a strange sense of distance: "Revolution is a necessity, not some kind of poetic inspiration. It's not about an age or a temperament, it's a damn necessity."

There is something tragic, in the deepest sense of the word, about these lives of great moral stature that nonetheless were experienced as the result of external forces, as a constant negotiation between living for oneself and living in the service of others. They seem like natural phenomena rather than acts driven by motivations, doubts, and personal conflicts. It is as though the actions of these revolutionaries have been completely subsumed by large-scale structural processes, which may be why their lives have only barely managed to generate a narrative of their own. Leninist heroes usually lack the psychological depth required by the modern novel; they seem more like Antigone than Madame Bovary. The socialist who appears in the middle of a traditional Cossack community in Mikhail Sholokhov's *And Quiet Flows the Don* is cold and distant, a ghost. There is neither goodness nor moral pride in this man who, nonetheless, has decided to dedicate his life to others.

Bertolt Brecht understood this perfectly well, and he fashioned not only his creative work but also his politics accordingly. Once more, Žižek captures this with great wit when he describes Brecht in the street in 1953, applauding the arrival of Soviet tanks headed toward Stalinallee to suppress a workers' strike. Likewise, in Walter Benjamin's lucid account:

> Marx, we may say, set himself the task of showing how the revolution could arise from its complete opposite, capitalism, without the need for any ethical change. Brecht has transposed the same problem onto the human plane: he wants the revolutionary to emerge from the base and selfish character devoid of any ethos.[4]

◎ ◎ ◎

The political movements that emerged from 1968's anti-institutional critiques maintained this objectivist tone. The year 1971 brought the famous debate on Dutch television between Noam Chomsky and Michel Foucault. While Chomsky took a fairly conventional Enlightenment stance—murder and oppression are bad, liberty and equality are good—Foucault offered a radically antimoral response that was theoretically cohesive but decidedly unusual:

> The proletariat doesn't wage war against the ruling class because it considers such a war to be just. . . . The proletariat makes war with the ruling class because, for the first time in history, it wants to take power. . . . When the proletariat takes power, it may be quite possible that the proletariat will exert toward the classes over which it has triumphed, a violent, dictatorial, and even bloody power. I can't see what objection one could make to this.

Twenty years later, Chomsky looked back on his encounter with Foucault:

> [I have never met anyone so entirely without morals.] Usually, when you talk to someone, you take for granted that you share some moral territory. . . . Usually, what you find is self-justification in terms of shared moral criteria; in that case, you can have an argument, you can pursue it, you can find out what's right and what's wrong about the position. With him, though, I felt like I was talking to someone who didn't inhabit the same moral universe. I mean, I liked him personally. It's just that I couldn't make sense of him. It's as if he was from a different species, or something.[5]

Why was the revolutionary Left so disinclined to examine its practices from a moral perspective? Why has it insisted on being "from a different species"? It is important not to give a condescending response. Several years ago, someone wrote in the Spanish daily newspaper *ABC* that "if the poor were more patient and the rich were more generous, everything would sort itself out." The phrase points to the shortcomings of consensus-oriented personalist ethics. Of course, not all theories are as antiquated and hypocritical as the *ABC* comment. The following statement appears in *The Short Summer of Anarchy*, a text by Hans Magnus Enzensberger about Buenaventura Durruti and Spanish anarchism in the 1930s: "In every town there was at least one socially conscious worker, who stood out because he didn't smoke, he didn't gamble, and he didn't drink, and because he was an atheist he never married his partner (though he was faithful to her)."

Socialism's resistance to relying too heavily on subjectivity has to do with the invisible disparities in orders of magnitude among social groups that appear to live in the same universe. In 1971, the Dutch economist Jan Pen came up with a way of representing social inequality to make its scale more intuitive. He called it the "Income Parade," a visual representation in which the income of each inhabitant of a country is represented by their height, in such a way that the poor are very short and the rich very tall. Next, we imagine that they all begin to walk past in a long line, starting with the shortest. The parade lasts exactly one hour. If we assume an average height of 5'6"—representing about $1,900 gross per month (the average in Spain for 2010, which is equivalent to about $1,450 after taxes)—the parade would go something like this:

It would begin with the very, very short, and the height of those marching past would gradually increase. After ten minutes, the people passing in front of us would be just over 3 feet tall (representing a gross monthly income of $1,100). Little by little, the height of those marching by increases, until at the half-hour mark—that is, the middle of the parade—the people passing by measure just under 5 feet ($1,675 gross). Five minutes later, we reach the average height of 5'6". To be honest, it's a pretty boring show. A whole lot of people pass by, and their height changes quite slowly. Around minute forty-eight, people who look like basketball players start to appear, measuring over 8 feet ($2,800 gross), and with about five minutes left, we see people almost 10 feet tall.

It's in the last minute that things finally get interesting. Extremely tall people begin to show up: the top 0.5 percent of the population, who measure more than 30 feet. These would include Spain's prime minister, Mariano Rajoy, who would be around 50 feet tall. Then come the few thousand salaried workers who earn more than $650,000 per year. First the shortest, who only measure about 160 feet (the length of an Olympic swimming pool), among them the former prime minister José María Aznar. Then come those with supersalaries, like Alfredo Sáez, the managing director of Banco Santander, who earns $10 million per year and would stand at just under half a mile, and the soccer player Cristiano Ronaldo, who makes $1 million per month and would be two-thirds of a mile tall. Even so, these are relatively modest heights compared to those of the super rich, who would pass by in a flash at the very end. In these cases, there are no salaries, of course. But if you take a fortune of some $1.5 billion with a yield of 4 percent per year, we would be looking at someone over 3 miles tall, taller than Mont Blanc. Even if we were to be more conservative in our estimates (say, a 2 percent yield), in those final moments of the parade we would see a truly unbelievable group pass by. They would include Amancio Ortega, the founding chairman of the Inditex Fashion Group, which owns Zara, among other brands. With an estimated fortune of more than $4 billion, he would be almost 40 miles tall and would have trouble breathing—his head would be in the mesosphere. Flipping the model around, if Florentino Pérez, the president of the Real Madrid soccer club, measured the average height, a normal person would look like a mite—that is, they would be invisible. If we took estates into account, the difference would be even greater, as would it if the parade were on a global scale. Roughly speaking, some 1,200 individuals in the entire world hold wealth in excess of a billion dollars, while a global population of seven billion people has an average income of $18,000 dollars per year.

What role do ethics play in the clash of the titans that is class struggle? The classic Marxist response is: not much of one—on this battlefield of fleas and giants, ethics are somehow absorbed by large-scale social relations. It is a little bit like that joke in which the pope goes to an African country suffering from famine and asks a cardinal from his entourage, "Why are these children so thin?" and the cardinal answers, "Your Holiness, it is because they eat nothing." The pope then bends down to one of them and says, gently, "But my child, you must eat . . ." Not even the most dogmatic among us would deny that our actions are at least a little morally vague. The appearance of structural inequalities on such a cataclysmic scale pushes this vagueness toward total semantic indeterminacy.

The resistance to the cult of personality in emancipatory traditions has to do with the idea that modernity is, in fact, a time of transition in which major social processes have a profound effect on our daily lives. As though we were experiencing a period of intense seismic activity in which the geological landscape was changing all the time. This has important ethical implications. For classical ethics, the social, cultural, and ideological context is just a backdrop, not all that different from the law of gravity. There are good reasons for this, from the perspective of argumentation: The emphasis on context is, in the end, a form of relativism. Furthermore, the economic and social conditions of most of these earlier societies were extraordinarily stable. In contrast, modernity is characterized by what Rousseau has called the "social maelstrom"—whether we realize it or not, major social change is a constant presence in our moral lives.

In general, demographic and economic shifts on the macro level produce significant ethical consternation. We suffer from cognitive limitations that keep us from taking on anything beyond a certain order of magnitude. This is why we all know that the representations of the solar system and of atoms are highly stylized images, not scale models. The distance between the planets is too big, and a subatomic particle too small, to be understood intuitively. For example, if we were to draw Earth the size of a tennis ball, the corresponding image of the sun would be more than thirty feet across. If we were to represent the nucleus of an atom with a peppercorn, a scale rendering of the atom itself would span sixty miles. It is as though we have a cognitive bias that affects our moral understanding of those actions that form part of complex, large-scale, and long-term processes. The Left has tried, perhaps unsuccessfully, to rise to this challenge.

There is a connection between this ethical asthenia and the illusions we hold today about overcoming traditional political frameworks through new forms of social relations. Perhaps this is

why cyberutopianism has taken such firm root among oppositional movements. The fetishization of the Internet takes the great
modern conflicts out of the equation and, in so doing, aims to turn
a major problem into a solution. The Californian creed has de-
prived the central revolutionary dilemma of its tragic dimension,
which was related to massive political and material conflict. It has
simply accepted this problem with digital cheer. It is a bit like that
episode of *The Simpsons* where Bart runs for president of his ele-
mentary-school class. Bart's adversary, one of the most diligent
students in their class, tries to run a smear campaign by putting up
posters that read: "A Vote for Bart Is a Vote for Anarchy." Bart
counters with a positive campaign of his own: "A Vote for Bart Is a
Vote for Anarchy!" Cyberfetishism cannot create ethical commitment, it's true. But in a time when machines define our social relations and no one talks about class struggle, this seems like good
news. One step closer to a frictionless society.

◎ ◎ ◎

Emancipatory movements have a theory about the nature of some
of the deepest social processes of our time. Essentially, they argue
that there is a profound connection between two historical dynamics: the Industrial Revolution and political emancipation. The idea
is that fully understanding one is impossible without understanding the other. There is no real political freedom or cultural progress if these lack a material dimension. As they say, *freedom of the
press is guaranteed only to those who own one.* By the same token,
the improvement of material conditions becomes an entropic process if there is no real way to act upon it politically.

  The basic revolutionary argument is that at some point in early
modernity the balance and positive feedback loop between political liberation and economic progress was lost. The range of political
options began to be geared toward the generation of economic gain,
limiting social development and generating processes of exclusion
and delegitimization. Material progress was also short-circuited by
overproduction, unemployment, financialization, and a tendency
to push the planet's ecological limits. All this was thought to have

caused not only the shortcomings of these political processes but also the break between the morality of our selves and our universal ethical theories.

Marxists tend to name the failure of the revolutions of 1848 as the beginning of politics' surrender to economics. That was the year the economy swallowed up all hope for the democratization and independence of public life. As Marx himself put it, the modern state is nothing but a board of directors that deals with the problems of the bourgeoisie. This is partly true and partly a figure of speech. To talk about Paraguay, where there is no income tax, and Norway as though they were the same is strange, to say the least.

In ostensibly more precise terms, it is said that the economy determines the range of political possibilities. In other words, economic processes do not exactly dictate what political organizations can do, but they do establish a framework that limits the choices these organizations can make. I generally agree with this premise, but I still have my doubts. As I suggested earlier with regard to the idea of causality in the social sciences, it is a fairly vague thesis. The limits or, more precisely, the opportunities do not automatically tell us anything about the choices themselves. Levine, Sober, and Wright offer an illuminating example.

> Imagine the following case: an individual chooses a pear from a basket of fruit. Two causes are involved: the range of fruits available in the basket and the person's preferences for different kinds of fruit. Suppose that there are thirty kinds of fruit in the world, and that twenty-five of them are included in the basket. Which is the more important cause of the individual's choice of a pear— the composition of the fruit basket or the individual's tastes in fruit? The answer is indeterminate given the information so far specified. It might be that, even if all thirty kinds of fruit had been available, the individual would still have chosen a pear. In this case, the structural limitation on the individual's choice is irrelevant. On the other hand, if the individual would have preferred one of the five excluded kinds of fruit, the limiting process would provide an important part of the explanation of the final outcome. In general, there is no simple way to determine

whether the reduction of possibilities represented by "limits" is larger or smaller than the reduction represented by "selections."[6]

This is not just an academic matter. If the range of choices permitted by capitalism were coextensive with the options offered by emancipatory projects, it would be hard to tell exactly how capitalism is the oppressor. This is an argument that many find convincing—at least, many people living in First World welfare states during periods of economic growth.

An interesting counterargument is that, as we know, the opportunities presented to us meaningfully influence our desires: We tend to want what we are able to get. As such, these limitations might keep us from really knowing what we want. If opportunities do not abound, we cannot be sure that we are making our own decisions, even when we think we are. Just like in *The Matrix*, we choose the blue pill.

Yet it is also true that we tend, collectively, toward a lack of willpower. If we acknowledge that we systematically tend to make self-destructive choices, we might conclude that being somewhat shackled might in fact be liberating—or, at least, might be a good plan B. Some individuals with problems controlling themselves stick their credit card in a glass of water that they then freeze, forcing themselves to pause before making purchases in the future. (Apparently, you can't put a credit card in the microwave without damaging the magnetic strip.)

This line of reasoning is often used to criticize some of the options presented to us by capitalism, such as privatization. We encourage restrictions on certain commercial options, even those that might be beneficial in the short term—by allowing, for example, a broader range of choices—because they would set in motion processes that we fear might become unmanageable or catastrophic. Nonetheless, as Montesquieu asserted, this argument could be used to defend opposing positions, as well. Capitalism could constitute a levee against other, less desirable possibilities, including the calamity of starting projects that are so virtuous they are impossible actually to bring to fruition.

I believe that the levee argument is misleading. All mechanisms of self-control involve some kind of review clause. When Ulysses asked to be tied to the mast of his ship in order to avoid the temptation of the Sirens, he was not giving up his autonomy for good; it was a limited agreement. People who freeze their credit cards have not been declared legally incompetent: They can still make purchases if they wish—they simply have to wait a few hours.

Our enlightened societies are resistant to irreversibility. This is why the death penalty has gotten such a bad rap and why we accept certain terms of employment that are worse than slavery but are not slavery itself. It is the mark of capitalism to seem like a collective form of self-control, when in fact it lacks a review clause. It is not, then, a mechanism of self-control but is instead subject to a law outside itself, one that has more in common with selling oneself into slavery than with putting a lock on the refrigerator. Marx had something along these lines in mind when he explained the role that the outward show of freedom and equality plays in the stratification of modern societies. Today, we permit unprecedented levels of material inequality because they coexist with respect for individual rights and equality before the law.

Socialism's rejection of morals is an attempt to match the asymmetry between the magnitude of the social and material forces that condition our actions and our capacity for ethical intervention. Certain social processes affect us so profoundly that they keep us from living our lives according to the moral principles we want to observe. It is what Günther Anders has called the "Promethean discrepancy"—the idea that we are now technologically able to produce major effects from minor actions. Innocent acts connect us to inconceivable repercussions: The mere fact of using our cell phone turns us into unintentional accomplices to the loss of thousands of lives in the violence that surrounds coltan mining.

A truly ethical position would be to choose freezing to death over wearing clothes produced by workers earning starvation wages. Revolutionaries are quite reasonably skeptical about the possibility of this kind of voluntary virtue catching on and have therefore given up on interpreting their own actions in terms of universal ethics. In this sense, the choice not to explain actions in moral terms could be understood as an attempt, though perhaps not a realistic one, to create a viable ethical framework. Anticapitalists want to build a society in which one can be good without being a hero and in which capitalist structures do not interfere constantly in our ethical, political, and aesthetic decisions. Brecht expressed this idea with great sensitivity in his poem "To Those Born Later."[7]

The theoretical amorality of the revolution situates itself, strangely, between two major ethical traditions. In principle, there is at least some overlap between socialist positions and the philosophical currents we might describe as contractualism. This would make the revolution a kind of "precontractualist" initiative, a foundational moment that seeks to establish the material and political conditions under which the social contract would not simply be an ideal or, worse yet, a fiction spun to legitimize injustice.

Contractualist theories try to imagine how a society should be organized in order to be considered fair and equal by every rational being within it—or, at least, by a sufficient number of Western rational beings. It is a way of understanding justice as a combination of rights and responsibilities established not by one particular group with its own particular perspective—not even the most saintly or noble individuals—but rather from an intersubjective point of view that anyone acting in good faith would accept.

For the philosopher John Rawls, the path to organizing a fair society involves imagining what principles people who will form part of that society would adhere to if they did not know what social position they will occupy within it. If I don't know which of the pieces of the pie I am cutting will end up as mine, the smartest course of action is to make them all the same size. In this sense, contractualism is the alternative to identity-based shortsightedness, as we see in one joke about architects. An ocean liner is about to sink, and the captain shouts, "Abandon ship! Women and children to the lifeboats! Men, grab a life vest!" He then sees a group of people on deck who haven't moved. "You! What are you all doing, just standing there?" To which one of them responds, "You didn't say anything about architects."

In effect, contractualist notions of justice say nothing about how each person should live their life: as a Catholic, a fashionista, a feminist, a soldier, an athlete, a rational egoist, or even an architect. Contractualism simply establishes a few abstract limits that allow some of these personal projects—the more, the better—to develop without restricting others. In outrageously general terms, from this perspective a fair society would be one that produces the most agreement among its members while at the same time allowing for the greatest number of different expressions of individuality. In other words, it tries to connect some version of the universality of moral responsibility—the idea that each person is affected by certain obligations that transcend context—and personal freedom.

The nexus of these two vectors is autonomy, the way in which people recognize one another as rational beings worthy of respect and as able to make decisions for themselves. Marxists believed that not only the Enlightenment but also capitalism fostered this kind of autonomy. Though capitalism significantly limits our political freedom, at least it freed us from an equally viscid subjection to external rules in the form of traditional relations of interdependence. The unfortunate prologue of the bourgeoisie has allowed us to conceive of a form of emancipation based on personal independence within a political body grounded in the consensus of free individuals.

Many postmodern philosophers and sociologists took this one step further and claimed that the capitalism of the information age was, in itself, an ideal setting for personal development. From their point of view, we live in an intense, exciting time in which we can all freely choose the right life path for ourselves without committing to that path beyond the duration of our interest in it. Not only are there no hegemonic notions of what the good life might be, in the sense that the social contract is limited to establishing a minimal framework for coexistence that guarantees the greatest degree of personal freedom possible, but the very concept of the good life has been taken apart, broken down into a succession of isolated preferences. The underlying idea is that both our personal identities and our societies lack a stable structure—and that this is a good thing. This is why the postmoderns were so quick to see a promising, imminent future in the Internet.

Socialism has little in common with this moral fragmentation. In fact, from another point of view, postrevolutionary society seems to adhere to an ethical model quite unlike contractualism. Oppositional movements have been somewhat, though not completely, vague when it comes to giving a detailed account of postcapitalist society. Marx said on occasion that in communist society individual development would occur freely and collectively, along with personal and creative growth. That is, socialism is not just a general framework in which people are free to come together to try to arrive at their idea of the good life but is, rather, a substantive ethical proposal.

Postcapitalism seeks to overcome bourgeois alienation and foster collective personal development. The Marxist notion of personal development, according to Jon Elster's elegant description, has to do with those activities that have an increasing marginal utility and are performed collectively.[8] Many of the things we like to do, like eating hot dogs, have a diminishing marginal utility: Every hot dog I eat brings me a little less pleasure than the one before it. The same thing is true of most goods in our consumer culture. In contrast, some activities provide greater pleasure the more they are performed. In a sense, they are an end unto themselves, which is why Aristotle called them "perfect acts." This is the case with music: Learning to play an instrument is an arduous task, but once you pass that first stage, each time you play is increasingly gratifying. Each book I read, or at least some of them, changes me in a way that no shirt that will be out of style in three months possibly could. The same is true of certain sports, political or artistic activities, and raising a child. Furthermore, some of these activities can only be carried out with other people, like playing Beethoven's *Pastoral*, participating in a municipal assembly, or making democratic decisions in a cooperative about a new production line.

In other words, socialism offers at least an outline of a better form of social organization. One does not start a revolution, however, in order to docilely accept the idea of a perfect life filled with Manolo Blahniks, paintball, and Disney cruises; for this reason, many postcapitalist projects maintain close ties with ethical theories of virtue. From this roughly Aristotelian moral perspective, ethics is a matter of making a good life in the context of the norms of a community and not just of seeking a broad or even universal social contract that establishes a framework for reasonable coexistence.

For Aristotle's heirs, the problem with the theories of liberalism is that they admit many projects that no one is in the position, socially or materially, to undertake. The case of the Internet illustrates this point. Though digital cooperation is not hindered by any obstacle related to the ownership of the means of production, it nonetheless remains a marginal practice because the institutional environment it requires does not exist:

> According to this communitarian objection, the liberal sees society as nothing more than a cooperative venture for the pursuit of individual advantage, as an essentially private association formed by individuals whose essential interests are defined independently of, and in a sense prior to, the community of which they are members. Conceptions of the good that are more strongly communal in content, that have as part of their very nature an insistence that social bonds are valuable in themselves, over and above their value as means to the attainment of other, merely individual, goods, are thereby downgraded.[9]

It is a bit like that Monty Python gag about a Protestant couple watching a long stream of children leave the house across the street. They are the offspring of a poor Catholic couple, who, unable to support them, have decided to sell them off as human guinea pigs:

> *Husband:* Look at them bloody Catholics, filling the bloody world with bloody people they can't afford to bloody feed!
> *Wife:* What are we, dear?
> *Husband:* Protestant! And fiercely proud of it!
> *Wife:* But why do they have so many children?
> *Husband:* Because every time they have sexual intercourse, they have to have a baby.

Wife: But it's the same with us, Harry.

Husband: What do you mean?

Wife: Well, I mean, we've got two children, and we've had sexual intercourse twice.

Husband: That's not the point. We could have it any time we wanted.

From something like a communitarian point of view, contractualism pulls the rug out from under its own feet. Its understanding of the conditions that would enable a fair social order eliminates the influence of individual definitions of the good life. In this way, however, it also prevents shared projects of personal development because once we take a formalist stance, we are condemned to seeing ethical programs as arbitrary, a matter of personal preference.

We cannot simply rebuild social ties when it is in our interest to do so and according to our own preferences. In the first place, communities are fragile. Social norms are easy to destroy and very hard to reestablish, and social relations are more like crystal than Play-Doh, as shown in a case study presented by the psychologist Dan Ariely.

A daycare center in Israel decided to impose fines on parents who arrived late to pick up their children. The initiative produced unexpected results:

Before the fine was introduced, the teachers and parents had a social contract, with social norms about being late. Thus, if parents were late—as they occasionally were—they felt guilty about it—and their guilt compelled them to be more prompt in picking up their kids in the future. . . . But once the fine was imposed, the day care center had inadvertently replaced the social norms with market norms. Now that the parents were *paying* for their tardiness, they interpreted the situation in terms of market norms. In other words, since they were being fined, they could decide for themselves whether to be late or not, and they frequently chose to be late. Needless to say, this was not what the day care center intended. . . . The most interesting part occurred a few weeks later, when the day care center removed the fine.

Now the center was back to the social norm. Would the parents also return to the social norm? Would their guilt return as well? Not at all. Once the fine was removed, the behavior of the parents didn't change. They continued to pick their kids up late. In fact, when the fine was removed, there was a slight increase in the number of tardy pickups.[10]

The reason behind this strange situation is that, second, social relations are more fluid than they are static. They are the unintended byproduct of complex processes that are extremely hard to reproduce.

◉ ◉ ◉

There is a special place in hell reserved for those who think they can bring together Kant and Aristotle, contractualism and the ethics of virtue. These are two opposing ethical systems, complementary in one sense but also contradictory. This is why the socialists tried to use the fantasy of the New Man to bridge the gap between the two. When we finally find ourselves in a society where we can establish a rational social contract without spurious economic interventions, a better version of the human being will appear, ready to take on the socialist way of life.

The New Man was a folksy name for the infinite flexibility of human nature, another of Marxism's great myths. Many socialists believed that our lives are completely determined by historical context and that there is no such thing as an atemporal anthropological organization. The arrival of a society of fair, happy, attractive, cooperative, altruistic, and satisfied individuals would depend on finding the right cocktail of social, political, and material relations. It was a heroic project. Cyberfetishism, on the other hand, seems to be a frivolous way of avoiding the problem by trusting electronics companies like Best Buy to provide technological sutures to mend these ethical options.

As a moral and social experiment, the New Man was a catastrophic failure, but it points us in an interesting direction. This failure forces us to consider precisely the opposite idea: that our anthropological characteristics, our "human nature," to use a polemic term, is relevant both ethically and politically to emancipatory projects. If we abandon the rather unreasonable idea that we are no more than clay that can be molded by society, our characteristics as a species take on political weight.

Modern ethics has not been very attuned to the major traits of humankind because it seems that, by introducing these matters into our moral reasoning, we risk falling into the trap of a naturalistic fallacy (ethical judgments cannot be based on matters of fact). By addressing only the rationality of a system of rights and responsibilities, it would seem as though we were not engaging any of these factual matters but were instead keeping to formal questions like consistency and cohesiveness. This sounds quite comforting, except for the fact that we are not mathematical sets. We are members of an animal species able to establish lasting bonds of kinship, assert our individuality, organize ourselves politically, create aesthetic and intellectual works, adhere to social norms, maintain personal relationships, perform economic exchanges, and so on. We can also do terrible things, like kill one another, and trivial things, like tickle one another.

An evaluation of a social system cannot take a neutral stance regarding whether that system is a suitable way of fostering those human abilities that form part of what we consider a "good life" in a broad sense. At least, that is what Marx's heirs have argued. By the same token, most of us—basically, everyone but the economists— have a hard time being immersed in a constant flow of competitive exchanges, feeling satisfied in a system marked by profound social inequality, orienting ourselves in society without stable interpersonal connections, and overcoming the deep-seated irrational biases that affect the decisions we make. Any political project that ignores these enduring realities could only be described as utopian, in the most negative sense of the word.

As a species we have not only great potential but also weaknesses. Behind formalist approaches to social organization is the idea that it is possible to develop procedures that would allow a group of rational and autonomous individuals to agree on the kind of social structure in which they prefer to live. It is an admirable idea, but the reality is that personal independence is, at most, a fleeting and not necessarily positive episode in most people's lives. Our rationality is affected by our vulnerability. We are animals plagued by problems, maladies, illnesses, and disabilities. We simply cannot live without the help of others. I do not mean this in the sense of a group of healthy, lucid, strapping adults coming together to prosper in hostile surroundings, as in the classical contractualist fairytales or the cyberutopias of today. From this perspective, disabilities are things that happen to us, when in fact they are a part of us: Children, for example, spend many years depending on those who care for them. Many people go back to depending on others at some point in their lives, whether sporadically or permanently. Put another way, we are beings that depend on one another, and any concept of personal freedom as the foundation of ethics must be consistent with this anthropological reality.

The philosopher Alisdair MacIntyre believes that human interdependence profoundly affects how our morality and rationality develop. The image of the kind of people we hope to become is closely tied to our participation in a community of abilities and obligations. Parents teach their children to postpone or adjust their immediate desires and to take on commitments and responsibilities. Later on, our peer groups and the people closest to us influence us deeply when it comes to developing moral positions.[11]

Social psychologists sometimes interpret these types of relationships in terms of submission to authority or to a group. Stanley Milgram's widely cited study demonstrated that just about anyone would be willing to perform an atrocious act if it was presented as an order from someone whose authority they respected. In the experiment, a scientist ordered the subject to administer increasingly intense electric shocks to an individual tied to a chair. Many obeyed, even when they believed they were inflicting deadly harm on the victim (who was, in fact, an actor and never in danger at all). Nonetheless, such experiments do not say anything especially negative about the role interdependence plays in our character as ethical subjects. On the contrary, they should instead make us want to avoid the hierarchical relationships so prevalent in bureaucratic and "total" institutions, which pervert interdependence and turn it into a source of moral aberrations. In fact, it is rarely pointed out that many of the test subjects of this experiment found the experience to be extremely useful in their own moral development. In the words of one test subject: "The experiments . . . caused me to re-evaluate my life. They caused me to confront my own compliance and really struggle with it. . . . I felt my own moral weakness and I was appalled, so I went to the ethical gym."[12]

The ways in which others depend on us are part of our ethical education. In many traditional cultures, older siblings play a crucial role in raising the younger members of the family. Parents raise their children until they are two or three, at which point they begin to depend on the older children, who are often only around ten.[13] Raising children is not a unidirectional service that independent individuals offer dependent ones; instead, it is part of the development of the older children. In order to become an adult, you fulfill the responsibilities involved in raising a child.

Our abilities, disabilities, and inabilities are all constant presences in our lives. The likelihood of our accomplishing anything, either individually or collectively, is inextricably tied to how we help one another. If we do not consider our natural inclination toward interdependence as a political issue, we will have no reason to think politically about the way we should depend on one another. The answer would already be a given: We shouldn't depend on one another—at least, not ideally.

In our societies, disability and dependence are seen as justifications of heteronomy: They are a phase either in the process of becoming an autonomous individual (in the case of children) or of no longer being one (for the elderly), or they are the result of surviving a catastrophe (such as with the disabled). Autonomy is a virtue reserved for the few (basically, rich white men in good health). Hence the long and appalling list of groups affected by controlling paternalism in the modern age: women, the poor, unskilled laborers, non-Westerners in general, the socially marginalized, the illiterate, immigrants, the mentally ill, and so on.

We tend to think about dependency the same way liberals think about equality. They do not necessarily see it as a bad thing, but for them it is not an obligation or a permanent situation. In any event, it is a starting point for personal freedom. Liberals find it logical that all children should have the same opportunities but that there is nothing unreasonable about rewarding different talents unequally. They do not see inequality as being inherently degrading. True egalitarianism, on the other hand, posits that certain levels of inequality are aberrations that keep us all from leading good lives, regardless of how badly off the worst off are or what our own personal situation is.

From a contractualist perspective, meaningful cooperation is always somewhat paradoxical in that it is both a necessity and a choice. As long as you respect the general framework of rights and responsibilities, cooperation is something you can elect to participate in or not. On the other hand, if we see ourselves as fragile beings that depend on one another, we must think of cooperation as a human trait as essential as rational thought, perhaps even more so. Our lives are unthinkable without the care we offer one another. Imagining a scenario of universal instrumentalism is just as wrongheaded as imagining one of universal irrationality and mutual deceit. Not all important social relationships involve caregiving, but caregiving is the material foundation on which they all are built. Political communities, even when they are based on contractual fictions, are built on a network of interdependence. The setting in which we manage to overcome alienation, or not, is an impulse at the very core of our nature: taking care of one another.

Most of us have experienced this in the domestic sphere, more than anything because generalized commoditization has driven it out of other contexts, especially the professional arena. This is why some people feel that thinking politically about caregiving is like thinking about society as though it were one big family, as though we had to treat one another like siblings or cousins instead of autonomous citizens united by a common goal. Exactly the opposite is true. Taking care of one another is the material foundation of a rational political bond not tied to individual whims or contractual formalities. In this sense, recognizing the importance of caregiving is essential to overcoming relations of dependence that alienate and oppress, including the ones we have with our relatives.

One good reason to be wary of communitarianism is the way many traditional societies have transformed caregiving into paternalism and domination. But there is an even better reason to be wary of capitalism: the way it has destroyed the social foundations of interdependence by initiating its socially toxic, nihilist project. Cyberfetishism gives this program of social destruction a fresh coat of paint, making it seem friendly and appealing rather than apocalyptic. It speaks to us of digital communities and enhanced connections, but it is profoundly incompatible with the practice of caregiving, the material base of our empirical social bonds.

There are different types of communities, with different objectives and forms of participation. In the first place, there is a basic difference between exclusive groups, like a swanky country club or a traditional guild, and more open ones, like a religious congregation or a modern labor union. Both kinds of organization can be liberating and egalitarian or oppressive and elitist. I believe that the ethical dimension of all these forms of social commitment, of this interdependence, is the experience of caregiving.

In contrast to what I called the fundamental ethical dilemma of the Left, we face no paradox in the question of care. It is hard to find a happy medium between modern individualism and traditional collectivism because social atomization has a strong self-destructive component: It breaks down the social networks in which it takes root, as demonstrated by the anecdote about the Israeli daycare center. This is why it is impossible to resolve the issue along a continuum between individualism and collectivism: Not even a small dose of rational egoism is compatible with a tightly woven social fabric.

On the other hand, we know of different experiences of caregiving, experiences that run the ethical gamut from domination to individual freedom. We do not have to choose between these opposing, mutually exclusive extremes. Caring for someone or being cared for is not a form of dominance or submission in and of itself but rather a part of our nature as deeply rooted as our ability to communicate and express fondness for one another. It may lead to violent and unequal power relations, but it does not need to do so—and in many cases it does not, in the least. The modern world has made an unprincipled effort to ignore this anthropological reality and to replace it with connections not grounded in interdependence: bureaucracy, information technology, commerce, ideology. The results have not been good.

The sociality offered by capitalism may be thriving, but it is always extremely superficial. Such is the case of the reticular social bonds in postmodern societies. The Internet is full of social relationships, but these prove useless when it comes to caregiving. Our family and friends, even our neighbors, are slow and annoying, but they are also consistent and reliable. The exact opposite is true of the digital sphere. The Internet is great for sharing television series but not for offering care. The fantasy that the former is as important as the latter is typical of individuals who have pathologically extended their adolescence and believe that online gaming is an intellectually and socially satisfying experience. If we've learned one thing from capitalism, it should be that alienation and a lack of solidarity are perfectly in keeping with a good quality of life and high standards for education.

Interdependence has nothing to do with connectivity. In a certain sense, they are antonyms. Capitalism is compatible with reticular social relations and trivial cooperation but not with caregiving. This is why there are still so many caregiving jobs outside the relationships defined by the market but that are nonetheless indispensable to its proper functioning.

This means that capitalism maintains a parasitic relationship to caregiving. According to some estimates, the labor represented by unpaid caregiving is roughly 50 percent of a developed country's GDP. Every morning, a vast army of workers and consumers rises from its beds—fed, bathed, healthy, and ready to keep running in the great hamster wheel of capitalism. Without the unpaid labor of caregiving, this would be simply impossible. All sophisticated notions about our place in a global network of dynamic and ethereal connections collapse when faced with the brutal physicality of caring for a newborn or looking in on a sick friend. There are few experiences that make us more aware of the limits of the modern economic system as trying to reconcile a typical salaried job with the care of someone who needs help. Interdependence, however, is not limited to these extreme cases. It affects us all to one degree or another.

The view favoring interdependence is consistent with the Left's mistrust of political ideologies that demand that any plan for social change respect the established political framework. It is also consistent with a profound criticism of how the pursuit of profit as a social motor systematically hinders all attempts to improve the lives of the majority. One striking statistic is that trust in others is highly correlated with material equality, at least in developed countries. Beyond a certain economic threshold—that of the OECD countries, to be clear—each relative increase in inequality whittles away at community ties, regardless of how wealthy the society is.[14] One excellent reason to defend egalitarianism is that inequality keeps us from caring for one another, robbing us of a chance for personal development.

The ethics of caregiving explicitly connects the kind of people we should aspire to be—the ideal of a good life—with the kind of social relationships we can hope to attain as rational interdependent animals, and it reveals the incompatibility of these relationships with the fundamental characteristics of capitalism, like material inequality and individualism. In this sense, it not only questions cyberutopianism but also allows postcapitalist projects to reconnect with their moral traditions, which they themselves have tragically insisted on denying. The organizations of the Left not only came up with an alternative to capitalism (albeit one that is not always reasonable or appealing); they also boast an interesting history of cooperation characterized by levels of commitment practically without equal in the modern age.

In his critique of Internet-centrism, Morozov reminds us of Kierkegaard's position on the media of mass communication that were emerging in the first half of the nineteenth century.[15] While most thinkers of the time were celebrating the rise of the mainstream press as a road to democratization, Kierkegaard believed that it was to the detriment of political life. Newspapers at the margins of the structures of power made it possible for their readers to hold passionate opinions about practically any topic of public interest. However, they did not develop the same zeal for acting on those ideas. Quite the opposite, in fact: The saturation of opinions and opposing information lent itself to the indefinite postponement of important decisions. The press unquestionably destroyed real political activity, which for Kierkegaard had to do with intense commitment and courageous choices.

The revolutionary organizations of the twentieth century seem to have taken Kierkegaard quite seriously. Regardless of whether you find the adventures and persecution of Leninist career revolutionaries appealing, these individuals were undeniably courageous, in a very literal sense, and committed. These same activists, however, were not exactly known for taking care of others. Ulrike Meinhof, for example, tried to send her children to an orphanage in Palestine. There is something paradoxical about this, given that these individuals developed a deep and freely chosen sense of commitment that the rest of the modern world only really experiences in terms of family relationships. Only a few very wise revolutionaries understood the parallel between the two things. As one anarchist recalls of a 1936 encounter with Durruti:

One afternoon we went to his house and found him in the kitchen. He was wearing an apron, and he was washing the dishes and cooking dinner for his wife and young daughter, Colette. The friend we had gone with tried to make a joke, saying, "Hey, Durruti—that's women's work." Durruti responded, curtly: "Take this as an example: when my wife goes to work, I clean the house, make the beds, and cook. I also bathe and dress my daughter. If you think that being an anarchist means sitting in some bar or café while your wife goes to work, you've missed the whole point."[16]

Durruti's statement is not like Simone de Beauvoir's assertion that no woman should be able to stay at home and raise her children, that no woman should have that choice.[17] Durruti understood that the work of caregiving should be seen as an important skill and part of personal development, not as a historical burden that men should lift from the shoulders of women.

Most organizations on the Left are blind to this truth. A few years ago, a neighborhood assembly of the antiausterity 15-M movement got into a debate over what time it should meet. Until then, they had been meeting at noon on Saturdays, but the summer heat was getting oppressive. There were quite a few parents of young children among the members, who suggested that ten in the morning might be a good time. The younger members without children protested, horrified: They went out on Friday nights, and there was no way they were getting up that early the next day. They thought it much better to meet at eight in the evening. The parents responded that at that time they were busy with baths, dinner, and story time. What surprised me was that the young members without children seemed to view caring for a child as just another lifestyle choice, one undeserving of any special consideration. Some people like going out and getting drunk on Friday nights; others prefer having children. You choose between beer and changing diapers the way you choose between MasterCard and Visa. End of story.

Many revolutionary organizations emerged as associations of mutual support that tried to mitigate capitalism's destructive effects on social relations. They seemed much more like soup kitchens than underground militant cells. This is a valuable institutional inheritance, which connects today's aspirations toward emancipation with nearly universal anthropological tendencies that have taken different organizational forms throughout history. For example, a common-pool resource is essentially an institutionalized system of mutual care based on commitment rather than empathy or solidarity. Similarly, the members of a traditional guild forged bonds of interdependence and reciprocity that are completely foreign to our understanding of professional relationships today. The idea of firing an incompetent apprentice would seem about as absurd to the master of a guild workshop as the idea of deporting an annoying nephew would seem to us today.

Oppositional movements tried to go a step further, questioning the hierarchies characteristic of old-fashioned communitarian systems and trying to filter out those oppressive vestiges. This was a mistake not because it was a bad idea but because they tried to do this by breaking with the ethics of caregiving and interdependence in favor of objectivity. This may be why the revolution has been more interested in the false promises of the social sciences than it has been in ethics. It may also be why the bureaucratization of care through a rationalist and impersonal system has been one of the great threats emancipatory movements have faced. I don't know if left-wing activism is the childhood illness of communism, but bureaucracy is most certainly its senile dementia.

Several years ago, the philosopher Carlos Fernández Liria told me that he thought socialism was an ideology for the weary, of people who need a break from the labor market, consumerism, advertising, and even from leisure. I think of it more like an ideology for tired parents and children. We should not trust any liberation movement that not only says nothing about interdependence, like most political programs today, but that in fact cannot say anything about it, as is the case with postmodern identity politics and cyberutopianism. Emancipation, equality, the free and shared fulfillment of our potential—none of these things can be separated from the collective care of our weaknesses. In a sense, that would be giving too much over to capitalism. Nonpaternalistic interdependence is the raw material we can use to design a set of friendly and egalitarian social institutions.

INSTITUTIONAL IMAGINATION

For years, social scientists have tried to describe the bonds within a community in abstract terms that were either strictly formal or not. The results have left much to be desired. The golden age of the social sciences is over. Though there has not been much criticism of its false promises, these days no one takes the musings of Talcott Parsons or Lévi-Strauss very seriously.

These zombie theories, nonetheless, still have an immense effect on our daily lives. The social sciences emerged in the nineteenth century as a theoretical means of confronting the practical problems of modernity. What we have inherited from this aspiration, which has been conceptually disengaged and transformed into a vague but ubiquitous worldview, are cyberutopianism and sociophobia. The ideological valorization of the digital sphere rests on the extraordinary technological developments of our societies, but it is grounded in a notion of our shared existence as a mere conceptual category defined by abstract qualities that unites fragile and fluid individuals.

The postmodern world has adapted itself to the false promises of the social sciences, not unlike the way a number of Raëlians committed suicide in order to bring about the prophecies in which they believed. It is as though everyone agreed to limit their perception of reality to the explanations provided by economists, sociologists, and psychologists. Digital sociality is so far-reaching because it is purely a formal relationship that is deemed correct before its material content is ever evaluated. The secret is that it barely has any content at all, as MacIntyre illustrates with this parable:

> There was once a man who aspired to be the author of the general theory of holes. When asked "What kind of hole—holes dug by children in the sand for amusement, holes dug by gardeners to plant lettuce seedlings, tank traps, holes made by roadmakers?" he would reply indignantly that he wished for a general theory that would explain all of these. He rejected ab initio the— as he saw it—pathetically common-sense view that of the digging of different kinds of holes there are quite different kinds of explanations to be given; why then he would ask do we have the concept of a hole? Lacking the explanations to which he originally aspired, he then fell to discovering statistically significant correlations; he found for example that there is a correlation between the aggregate hole-digging achievement of a society as measured, or at least one day to be measured, by economic techniques, and its degree of technological development. The United States surpasses both Paraguay and Upper Volta in hole-digging. He also discovered that war accelerates hole-digging; there are more holes in Vietnam than there were. These observations, he would always insist, were neutral and value-free.[18]

Social scientists often do little more than collect everyday concepts—which are necessarily vague and connected only by family resemblance, as in "hole"—and then elaborate vacuous theories that are nonetheless formally sophisticated and sometimes even erudite. Not only do these idiosyncratic theories require a formidable investment of time and energy to produce; they also influence public policy, inserting themselves in these debates by costly, morally ambiguous, and practically inefficient means.

Economic, sociological, political, pedagogical, and psychological theories have played a major role in some of the most important political transformations of the modern age. Social scientists have often been recruited to assist in organizing and regulating the justice system, the economy, labor relations, education, military strategy, and social services. Yet the various social theories employed have never been called to account for the pathetic results they have yielded, which seem clearly inferior to those which would have been obtained had a bit of common sense been applied or even if the status quo had remained undisturbed by this supposedly scientific knowledge. In a well-known informal experiment, the *Wall Street Journal* had a blindfolded monkey throw darts at a target covered in stock prices. The monkey's portfolio did better than those of 85 percent of the fund managers in the United States.

Indeed, economists have turned their field into a branch of applied mathematics remarkably far removed from notions of material subsistence, processes of production, and historical social relations. As the political scientist Peter Gowan has asserted, the accumulated knowledge of financial experts is often a barrier to understanding the economy. Specialists consistently engage in practices that are an affront to even the most basic sense of prudence. The systematic failure of these ideas has not dampened the vehemence with which they are defended; the fact that Karl Popper, a thinker obsessed with the verifiability of scientific theories, is just about the only philosopher of science whose work is read in economics departments only adds a layer of irony to the idealistic daydreams so often mistaken for mathematical rigor.

In real science, deductive reasoning yields solid empirical results because it taps into stable cores of intelligibility within the phenomena it seeks to explain. This is why physics allows us to work in a mathematical fashion with clearly defined quantities and obtain precise results, but the same is not true in the social sciences or economics. Our rational and irrational behaviors are particularly resistant to formal analysis. Of course, with enough patience, practically anything can be codified, even family or relationships of affinity. But in this pseudoscientific context, the operations performed with the figures thus obtained are empirically meaningless—pure speculation that only from a distance might be mistaken for mathematics.

The social sciences are praxeologies, just like translation, cooking, politics, literary criticism, raising children, playing sports, farming, and playing a musical instrument. In all these fields, there is knowledge and ignorance, a distance between doing it right and doing it wrong. They are forms of practical knowledge in which experience, absorbing and building on prior empirical data, imagination, and analytical development are key factors. The original sin of the social sciences is to extrapolate ideas from this everyday knowledge and to mobilize them as though they were scientific concepts in their own right. Science does not advance by merely putting commonsense notions into systematic order. To the contrary, it represents a break with our everyday experience.

Aristotle assigned the term *phronesis*, or "prudence," to the kind of practical knowledge we engage when we want to change things for the better, whether in our own lives or in public policy. The *phronimos*, or prudent person, is the one who will act most suitably in situations that cannot be reduced to general principles. Prudence is not theoretical knowledge regarding experience but is instead a form of wisdom that emerges through practice. Not a food critic, but a chef; not a pedagogue writing a textbook, but a teacher in the classroom; and so on. Phronesis gets a bad rap as an unsophisticated platitude that consists of finding the middle ground between extremes: "be neither stingy nor extravagant, brash nor cowardly . . ." In fact, the opposite is true. Phronesis addresses intense, often tragic, practical issues like how to behave in armed conflict or have a good relationship with a friend or one's child. The solution only seems obvious after we have found it, once the process has been carried out successfully. The only proof that we've come up with the solution to a practical problem is that it seems reasonable to us. When most people or the wisest among us find a way around an issue, the answer seems self-evident—but this was not necessarily so before the process of reflection that revealed it.

Cyberutopian postmodernity is so receptive to the abstraction of the social sciences because it lets us fool ourselves into thinking we've overcome the problems of modernity without this sort of tentative practical intelligence playing any part in the process. On the surface, the question of which political structure would allow us to overcome our most urgent public issues is practically rhetorical. The answer is none. Those who fetishize the Internet don't need their freedom to be collective—that is, shared—they only want it to be simultaneous. The Internet offers a superficial substitute for emancipation by alternating doses of independence and connectivity. The social metaphors associated with digital networks make political interventions driven by consensus appear slow, coarse, and boring in comparison to the spontaneous, organic vitality of the Web. The very design of the digital sphere encourages us to expect that the best solutions will emerge automatically, without the adjustments and corrections that come from debate.

Deep down, this is an unconscious reflex of an age-old ultraliberal ambition. Naomi Klein rightly said that Milton Friedman's worst enemy was not communism, which he considered to be merely wrongheaded, but rather Keynesianism.[19] He believed that Keynes had proposed an unclear, grotesque hybrid that refused to give up the game of supply and demand but allowed political institutions to distort it. Neoliberalism demands that we organize our existence around procedures that are as coherent and consistent as a well-crafted language, along the lines of a logical precept. From this extreme point of view, abstract properties are more important than material effects. Cyberfetishism and consumerism have fed into this bias from outside the economic sphere; as a result, those social relations that by nature do not correspond to this narrow definition—like taking care of one another, forming deep affective bonds, and developing a political stance of one's own—have become invisible.

Despite its limitations, the formalist spontaneity of social networks and digital connectivity seems like a good option to us because analog politics have proven shockingly ineffective when faced

with the power of the market. For example, in August 2011 one let-
ter from the president of the European Central Bank was enough
to blast through one of the biggest taboos in democratic Spain. For
decades, the Spanish Parliament had unanimously held that not
even the smallest modification could be made to the constitution.
The Spanish Constitution of 1978 was seen as the coda to a fundamental political process—the transition to democracy that carried
us out of Franco's dictatorship. It was thought that any interference
in its delicate gears would make the whole thing fall apart, throwing us into social backwardness and fratricide. Nonetheless, in just
a few days the government and opposition party secretly agreed
upon a modification to the constitution that established a ceiling
for public debt. Our very charter thus limits the nation's ability to
make decisions and instead favors the power of the market.

The only versions of collective sovereignty that we know today
are the result not of collective rationality but rather of atavistic or
religious influences or of identity politics. We see the Muslim world
as a collectivist magma—and therefore as fanatical and irrational.
Cyberfetishism and sociophobia are the final stages of grief in this
terminal case of modern heteronomy, in which we submit ourselves
to the market (and are no longer even angry or in denial about doing
so) and try to emulate its basic mechanisms in our social lives.

The greatest challenges to cyberfetishism and sociophobia are
not Luddites and communitarianism but rather politicization. One
can maintain the fantasy that social interactions within digital
networks might help us overcome poverty, the alienation of labor,
loneliness, or global warming, but this kind of antipolitical day-
dreaming is incompatible with detailed institutional planning.
Collective efforts aimed at fostering equality, caregiving, and personal and professional development need to assert democratic rule
over and against the heteronomy of commerce. Nor is it enough
simply to define these projects in abstract terms and then set them
in motion as though they were a neural network. They require a
sustained commitment to their correction and improvement, like
a literary translation or a common-pool resource.

For example, there was a time when microcredit schemes seemed
to be a vehicle for social transformation in poor countries. One of

the reasons behind their popularity was that they are a kind of economic version of digital cooperation. Microcredit seems like a reticular strategy that needs no centralized coordination, acting as an initial financial push that generates a spontaneous and autonomous form of empowerment. Ideally, microcredit provides economic tools to help families develop their own projects, without the need for new political institutions that would intervene in a meaningful or prolonged way. Nonetheless, 2012 saw a wave of suicides in India related to microcredit schemes, which brought to light how the initiative had created a financial bubble in which many then found themselves trapped. People living in extreme poverty who had applied for small loans took their lives when faced with the impossibility of paying back their debt. The explanation given by microcredit enthusiasts is that Muhammad Yunus's original project, which had a social objective, had been corrupted by speculative investments. True, but it is still revealing how unrealistic these projects of social transformation can be when they fail to take into account the institutional context in which they will be carried out—such as the absence of antiusury regulations. The experience is surprisingly similar to that of Nicholas Negroponte's hundred-dollar computer.

Oppositional politics occupy a strange space in this landscape today. The revolution developed conceptual tools that were useful in diagnosing the political limitations of capitalism, but it did not dare to do away with some of its false promises. In general, what socialism has said about the post-Neolithic context of production—that is, industrialized societies—is that capitalism is not equipped to manage it. Capitalism is inefficient, in the sense that it systematically misses the opportunities it generates. It has not been able to take advantage of its own historical potential. In other words, the point is not that capitalism is not the system best suited to developing a labor force (it may well be) but rather that its use of that force is, socially, far from ideal.

In broad terms, today's technologies should allow many people a higher standard of living without drastically lowering the standard of living for those who are better off. In such an alternative system, a few of the megarich would probably have to live without their yachts upholstered in whale-penis leather, the Japanese middle class might have to accept the idea that a life without automated toilets is one worth living, and residents of the United States might have to get over the idea that bicycle lanes are a sign of the coming apocalypse. On the other hand, around a billion fewer people might go hungry, and a similar number might learn to read and write. Given the ecological limitations of our planet, the days of North American consumerism are numbered, anyway: In most Western countries, sustainability is already synonymous with degrowth. Indeed, it is a proven fact that greater income equality produces benefits—in terms of quality of life, life expectancy, and a general reduction of various social problems—across all social strata, not only the most disadvantaged.

This argument may seem irrefutable, but it is more problematic than it appears. One day I was giving a class on Marxist theory and explained the matter of capitalist inefficiency by giving the example of the famous light bulb that has been illuminating a California fire station for over a hundred years. It seems that in the first decades of the twentieth century, the owners of the biggest light-bulb factories got together and agreed to artificially limit the life of their products to one thousand hours, though they could last much longer. This is a good example of how capitalism is unable to realize its full material and social potential because it is driven only by the profit motive. Raul Zelik, a professor of political science from Germany who was sitting in on the class, raised his hand and said, derisively: "Lovely story, but how do you explain the fact that in East Germany light bulbs used to last only 500 hours? And that no agreements were needed for that to be the case?"

The capitalist system may leave much to be desired, socially—but that doesn't mean there is a more efficient, viable system out there. We anticapitalists have decided, in a fairly uncritical way, that there is a social alternative able to make better use of the resources capitalism puts at stake. But what if there isn't? What if the best option is simply not available to human societies?

Classical socialism, Marx included, assumes that the carefully planned distribution of resources is more efficient than the chaos of the market. At first glance, capitalism's approach to satisfying social needs is like throwing thirty darts at a board and hoping one hits the bull's-eye. Competition over providing goods and services has led to squandering of epic proportions. One-third of the European Union's food is thought to end up in the dump; the 40 million tons of food thrown out every year in the United States is enough to feed the hungry all over the world.

We are, however, also well aware of the empirical difficulties involved in minimizing these inefficiencies by means of a centralized system. Strangely, reflections on the Soviet period and the problems with planned economies have become scarce since the nations of the Warsaw Pact made the transition to capitalism. While these countries were exercising "real socialism," they modernized their primary sectors quickly and with reasonable success. They were also able to provide complex social services like education and health care fairly effectively, but they failed with consumer goods and services. Of course, the political costs—oppression, violence, authoritarianism, and alienation—were high and widespread and should not, under any circumstance, be dismissed as a particularity of Slavic or Eastern socialism.

The typical response of the non-Soviet Left to the shortcomings of real socialism has been to blame bureaucracy and authoritarianism. It is true that bureaucracy tends to ritualize administrative procedures to the point of turning them into an end in themselves, which is clearly incompatible with a dynamic economy that demands flexible reactions to different situations. Furthermore, by specializing and codifying these managerial roles, a critical share of power is given over to administrative heads.

From this perspective, the answer to the longstanding problems that plague social planning is democracy: All that would be required for planning to work would be to tip the bureaucrats out of their armchairs and let the workers decide the details of production in town hall–style meetings. This is a pleasant but misguided idea. Bureaucracy is, in fact, a rational response to the enormous problem of organizing the entire economy of a complex society. Armies, for example, are relatively simple planned societies, and they are full of bureaucracy, even though this bureaucracy is not an essential part of their hierarchical systems.

The real problem with centralization has to do with the fact that it is not entirely clear we are even able to make optimal economic decisions if we are not motivated by their outcome. Providing goods and services in a complex economy involves a staggering number of microdecisions. Any deliberative process, whether authoritarian or democratic, implies a high degree of uncertainty because it cannot take all these variables into account. Alec Nove, a socialist economist critical of Soviet centralization, sums these problems up quite insightfully:

> One can issue an order—produce 200,000 pairs of shoes—and this is identifiable and enforceable. To say "produce good shoes that fit the customer's feet" is a much vaguer, non-enforceable order. (Similarly I can be meaningfully instructed to give fifty lectures, but it is not so easy to enforce an order that I give good lectures!) It also shows the severe limitation of planning in physical quantities. The same number of tonnes, metres, pairs, can be of very different use-values and fulfil widely different needs. In any event, quality is a concept frequently inseparable from use; thus a dress or a machine can be fully in accord with technological standards, but still not be suitable for a particular wearer or factory process. How can this problem be overcome if plans are orders of superior authority (the central planners or ministries) and not those placed by the users?[20]

It is not that the market is a particularly elegant alternative to centralization. Even in the best of all possible worlds, it is one big process of trial and error that wastes huge amounts of society's energies. What is not clear, however, is that planning could, even in theory, determine global supply and demand for goods and services in a complex society, much less come up with the models of production and organization necessary to satisfy this demand.

One technological response to the limitations of centralization has been automation. There is a long tradition of attempts to computerize socialist planning: This is the prehistory of cyberfetishism and digital collaboration. In summary, the idea is that if bureaucracy fails because it is clumsy, slow, and subject to human weaknesses like the hunger for power, it would make sense to replace bureaucrats with fast and ethically neutral machines. The history of computation in the USSR was closely linked to the development of technological tools that would facilitate the optimal distribution of available resources, minimizing bureaucratic interference.[21] Soviet economists were looking for some kind of digital substitute for the free market. In cybernetic planning, technological tools would replace pricing in its function of providing the information the system needs about the preferences of individual and collective actors in society in order to optimize outcomes.

Of course, there is nothing outlandish about thinking that computers might be of use in organizing economic structures. Nonetheless, the limitations of planning are less a product of calculation error than of a practical dilemma. The microcontexts that make up a complex economy are dynamic: They cannot be defined ahead of time, and those definitions, when formed, are qualitative in nature. The definition of a "good shoe," to stay with Nove's example, depends on the context and is hard to define a priori (shoes for hot weather, for mountainous terrain, for dry cold, for the rain, and so on). In the same way, the availability or absence of certain goods or services alters the preferences of the producers and the consumers.

Strangely, the gaps in computerized planning bear a remarkable resemblance to the most significant limitations of orthodox economics. Every microeconomics textbook opens with a model known as "perfect competition." The setting seems familiar: a market square in which buyers and sellers haggle over wares. In fact, perfect competition is exactly the opposite: a surreal scene quite similar to the centralized digital market that some Soviet economists tried to develop. Economists are most concerned with proposing sophisticated models, but they face major limitations in their attempts to capture, in mathematical terms, a truly competitive context. Of course, none of them would allow social reality to sully an elegant equation, even if it means having to describe capitalism as though it were some kind of five-year plan:

> The model of perfect competition begins by assuming that the agents are not authorized to set the prices of the goods they are

buying or selling. The prices are "given" from the outset. They are not the result of a process of negotiations or conversations among the members of a society . . . the authors of these textbooks know that the model of perfect competition describes a centralized system. But it is hard for them to accept this, which is why they never say it clearly. Sometimes they allude to an "auctioneer" who announces the prices that will serve as the basis for these exchanges, but they never include this figure in the indexes of their books, which are generally very detailed. It is as though they were ashamed of him.[22]

According to the economists, we are robots who only think we are free: We want as many goods as we are supposed to want, at the price we are supposed to be willing to pay for them, as though the market had been designed by a master planner. Deep down, this is not a bad description for the absurdity of actual capitalism. What neoclassical economics does is shift the problem of social planning to inside the heads of those who intervene in the market economy. As though, in our minds, the concept of "a good shoe" were some kind of perfectly established Platonic ideal. This is a particularly strange argument when it comes to goods and services that are either complex or that have significant externalities, like housing, transportation, and energy, which are dynamic by definition and seem to depend on collective reflection and social norms.

In 1950, for example, 60 percent of all travel within Spain was by train, while 40 percent was by car or bus. By the end of the 1990s, train travel had declined to 6 percent, while journeys by road had climbed above 90 percent. This change was not the result of decontextualized personal preference but rather of collusion between active and expensive choices in public policy (today, Spain is the European country with the most miles of highway per vehicle and resident) and the private interests of economic elites in the construction and automotive industries. Institutional support for private road transportation instead of other alternatives redefined the physical structure of our cities and transformed our idea of what makes a given mode of transportation effective. This is why we opt for a vehicle that is expensive, dirty, and shockingly slow (the average speed for a car in a big Spanish city is about ten miles per hour). Our need to own an automobile and the interest many people have in cars as status symbols are partly a result of dynamics that, if they had been put to collective discussion, might have been reversed. Each time we feel the need for a car, sixty years of political economics course through our libidinal circuitry.

For orthodox economists, the market acts like a hive mind that on the one hand renders any attempt to build consensus about collective preferences unnecessary and, on the other hand, allows us to overcome our limitations in order to find the best means of satisfying them. Public discussion is unnecessary. Collective decisions are the inadvertent byproduct of social interaction between individuals who do not coordinate among themselves. Through pricing, each individual can know what they need to know in order to organize their perfectly clear economic preferences.

There is no doubt that the atomization of decision making and the absence of collective deliberation greatly increase the danger that individual irrationalities are reinforced, creating a catastrophic collective snowball effect. We tend to call this historical avalanche "capitalism." The intuitively strange fiction of the auctioneer points to the fact that, in reality, there is no reason to think that the intersection of optimal individual decisions should give rise to a situation that is desirable for the majority. Adam Smith and the founders of eighteenth-century liberalism leaned on divine providence to hold onto the belief that this would be so. These days we have game theory—a science that rests on far shakier foundations than theology.

Individually speaking, we cannot really know what we want. Karl Polanyi has said that in a market society our preferences are unstructured. He meant this in the sense that we lack a framework of norms by which to orient our priorities. This is a good estimation of the ruin brought on by consumerism. Even when we try to be reasonable and put the satisfaction of our basic needs like shelter, food, and warm clothes ahead of our desire for luxury goods, we end up with a forty-year mortgage on a condo unit in the Mediterranean holiday city Marina d'Or, on the verge of morbid obesity from eating trans fats, while we walk around dressed in ridiculous, obscenely overpriced clothes.

In fact, it's even more complicated than this. We cannot make our preferences adhere to even a minimal standard of consistency. The value we assign things is intrinsically ambiguous—a diffuse idea poorly defined in our own heads. This is why the way we describe a given situation has such an effect on the choices we make. When some gas stations in the United States started charging customers who paid by credit card an extra fee, the customers called for a boycott. The response from the gas stations was to raise prices across the board and offer a discount to those who paid in cash. The boycott was called off.

As with online social relationships, the price we pay for a formalist understanding of the logic of preference is an abyssal lowering of the bar of rationality. If one is inclined to accept poverty, inequality, pollution, and poor education as reasonable outcomes of economic processes, it is hard to imagine a system of production that would not be capable of meeting such low standards. The only advantage of the market is that its shortcomings, which are not necessarily less significant than those of a centralized system, appear less obvious or less pressing. The fact that millions of people are left without sanitation because they lack the money to pay for it does not seem like the direct result of the free market in the same way a shortage of socks seems like the direct responsibility of the planner who should have anticipated that particular need. The free market gives us a set of blinders so we can ignore our practical limitations, whereas planning is a magnifying glass that amplifies them. Both proposals, however, are based on formalist false promises, the type of illusion that cyberfetishism has turned into an alienating utopian project.

In order to evaluate different institutional options without the distortions caused by aspiring to a system either entirely planned by the state or entirely governed by free trade, it helps to detoxify from the social sciences. Criticizing the false promises of the social sciences and their impact on our political life does not necessarily mean giving up the attempt to explain human experience or stating that we can only interpret it through literature. Yes, the human sciences are limited, but our immediate perception of society is too, and to an even greater extent. No, there are no theories of human reality, in the strict sense, but we can apply certain explanatory mechanisms to various social phenomena.

The term "mechanism" refers to a conditional kind of explanation that does not seek to generalize. Mechanisms are conceptual devices that can only be identified a posteriori, after an event involving one has taken place. They are causal but discontinuous explications that lack consistency and homogeneity, and as such they cannot be the source of a theory. Jon Elster provides an illuminating example:

> When people try to make up their mind whether to participate in a cooperative venture, such as cleaning up litter from the lawn or voting in a national election, they often look to see what others are doing. Some of them will think as follows: "If most others cooperate, I too should do my share, but if they don't I have no obligation to do so." Others will reason in exactly the opposite way; "If most others cooperate, there is no need for me to do so. If few others cooperate, my obligation to do so will be stronger." In fact, most individuals are subject to both of these psychic mechanisms, and it is hard to tell before the fact which will dominate.[23]

The same can be said of the political sphere that the social sciences have played a part in embalming. If we give up the false promises of the social sciences, perhaps we can undo the effect they have had on our political imagination. Categorical, unshakeable principles like individual freedom, democratic decision making, and material equality do not necessarily imply generalized institutional proposals, much less universal notions about social bonds. Radical political change is compatible with bets placed contingently on ideas for institutional transformation that are not overly formal or far-reaching.

Many socialists, authoritarian and otherwise, tried to destroy the material obstacles represented by capitalism but did not examine the practical limitations of an abstract understanding of political economics. There are some who believe that overcoming our market-driven society is simply a matter of redistributing profits that are currently held by only a few. The truth is that if we were to redistribute the annual earnings of the Spanish stock exchange when it was at its historical peak among all the citizens of the country, it would work out to about $750 per person. Most of us could probably use the money, but it's a far cry from egalitarian emancipation.

One indispensible anticapitalist tenet is that all public matters should be subject to democratic deliberation. This implies a subversion of the dominant tendency in liberal democracies to cut out of politics the debate over the accumulation of capital, as this accumulation is seen as a given. But continuing to discuss the matter is not the same as an overarching imperative to intervene. The problem with anything overarching is that it requires a high level of abstraction and pays little attention to context. Postcapitalism is not a guarantee that things will work out but rather that problems can be addressed without the interference of abstract allegiances.

This might appear shortsighted when compared to formalist false promises. The fact is, however, that the institutional architecture of any society is based on contingent substantive preferences. In real-life capitalism, support for the market is dependent on the financial gain of the dominant classes. When the market has not accomplished this aim in a satisfactory manner, it has been violently suspended from duty. This is why governments today consider nationalization an acceptable alternative when it is a matter of redistributing the inconceivable losses made by the banks. The formalist adherence to the surface of these dynamics is an ideological tool that exploits sociophobia and the discreet charm of apolitical spontaneity. Its greatest expression is the digital utopianism of today. The Left has often accepted these terms, believing that a democratic alternative to the chaos of the market would have to meet the same conceptual expectations. This is why cyberfetishism has had such an effect on oppositional movements.

Polanyi believed that different ways of institutionalizing economic relations—reciprocity, redistribution, the market, and the farmstead—actually always coexist. We establish relationships of reciprocity when we give birthday and Christmas gifts. We do not participate in these networks of exchange in order to reap any gain, although we tend to assume that some of the people whom we give gifts to will also give ones to us. Redistribution is the kind of structure exemplified by tax systems: A central institution gathers goods or services from different members and distributes them according to a preestablished norm. The market is a form of exchange based on bargaining and competition in which we participate in order to try to get ahead of others. Finally, the farmstead is a self-contained system that consumes what it produces.

There may be other ways of institutionalizing the economy, though there certainly aren't many of them. The real economy behind our material subsistence is a system of counterbalances among these different economic apparatuses. Political intervention can change this only so much, by promoting one institutional mode or another. Many traditional societies excluded goods related to basic human needs, like food and land, from commerce. Liberalism has historically been a system that requires constant interventions on the part of the state in order to avoid collapse. In the same way, there was informal commerce and a persistent black market in the Soviet Union. And of course, systems of reciprocity like unpaid caregiving survive in every community.

Alec Nove said that it is absurd to treat all goods and services within a complex society as though they were identical and subject to the same rules. He thought, for example, that it was completely reasonable to have a planned, centralized economy for goods in continuous demand—like water, energy, and transportation—while on the other hand he saw the market as an effective tool in producing the kind of goods and services that were neither in continuous demand nor a basic human need. Whether or not Nove was right, there is no question that commoditization tends desperately toward the homogenization of unrelated realities: financial transactions and food, labor and luxury cars, intellectual property and currency. Soviet centralization made the opposite mistake by thinking that the production of absolutely all goods and services could be effectively planned.

A postcapitalist society should be able to articulate its systems of production by means of institutions that vary based on the context. In this sense, it is critical to identify those economic options that set in motion self-destructive processes that are difficult to reverse, such as, for different reasons, the privatization of essential goods or authoritarian economic planning. There is no practical principle, however, of decontextualized organization. There is no reason for supporters of egalitarian, emancipatory alternatives to capitalism to come up with an entire postmarket system. They should instead be thinking about a social context in which the various economic institutions are subject to democratic decision making. As a result, in a system where the reach of economic mechanisms that threaten popular governance and its material foundations in caregiving is greatest—like income inequality and technocracy—it would have to impose meaningful limits, regardless even of its organizational efficacy.

It sounds unambitious, but this is the sad fog through which our practical rationality moves, in which any noncontingent principle beyond a few basic ideas about human nature is out of place. Perhaps this perspective is not so far removed from the original socialist program. Marx's refusal to give details about postcapitalist society tends to be understood in terms of a conceptual impossibility: Communism is so exotic, so radically different from our world, that we don't even have the vocabulary to describe it. But it could also be just the opposite: the choice not to present abstract arguments about what is simply the day-to-day reality of political immanence. It is interesting to note, though people rarely have, that *Capital* does not abound in praise for revolutionaries. On the other hand, in the prologue to the text, Marx describes factory inspectors—the Victorian equivalent of labor inspectors—as "competent, fair, and inflexible" men.

As such, perhaps we should also consider interpreting that famous though perennially ridiculed phrase from Marx and Engels—that through socialism we will overcome the alienation of labor and will be painters in the morning, teachers at midday, and doctors in the evening—in a different way. It may not be, after all, an absurd aspiration to create multitasking workaholics but rather a disavowal of applying homogenous categories to heterogeneous realities. Salaries create a formal equivalence between activities that have nothing to do with one another, some of which are creative and interesting, others of which are boring and excruciating. Opening up the institutional imagination means questioning this kind of homogenization and demanding that political decision making respect the contextual nature of our practical rationality. Thinking about postcapitalism means, for starters, refusing to call any bit of data found on the Internet "information," any remunerated activity "work," and any market-based choice the revelation of a "preference."

The basis of socialism has less to do with certain formal institutional principles—the rule of law, parliamentarianism, or general assemblies—than it does with modeling persistent human realities in a somewhat flexible way. One such reality is that the economy is not a domain insulated from the rest of social life but rather a slice of a sustained, collective relationship. Another is that we are interdependent, fragile, and only partially rational beings, not asocial angels that can survive on fragmentary, sporadic relationships. Also that recognizing one another as sovereign individuals is inseparable from the possibility of developing a good part of our human capabilities. And, of course, that material equality—and not only improving the situation of those worst off, or presenting equal opportunities—is an essential condition for social relations based on freedom and solidarity.

This is why institutional mechanisms always seem like a toolkit. They are instruments at the service of political deliberation that we may wish to apply in certain situations, but we cannot know in advance whether we will decide to implement them. Revolutionary institutional models have been lacking in this sense. An insistence on councils, democratic centralism, anarchosyndicalism, and collectivism have left little room to maneuver in relation to different situations and problems that any complex society undoubtedly experiences, like corruption, authoritarianism, or simple incompetence. These mechanisms have taken themselves to be abstract principles rather than contingent apparatuses.

Throughout history, solid and stable institutions have been open to the diversity of motivations and possible weaknesses of their members. The Catholic Church is exemplary in this regard (and surely only in this regard). For centuries, it has lived with avarice, brotherhood, authoritarianism, charity, submissiveness, cruelty, modesty, the will to power, venality, and the withdrawal from the world. Capitalism, on the other hand, is much less flexible. The attempt to turn competition, egoism, and fear into drivers of social behavior is not only amoral; it is also impractical. Capitalism is in permanent crisis and is incredibly fragile, above all when compared to systems of production that have been around for thousands of years. If it seems resilient, this is only because it produces extreme path dependence. Once you take that first step toward privatization and individualist confrontation, it is very hard to turn back.

In the 1920s, Richard H. Tawney underscored how the conflict between business owners and workers in capitalist industry prevented the honoring of responsibility, which was related to a social function, from being an important vector in the modern professional world. Professional life under capitalism, Tawney said, is organized around the defense of opposing rights—those of the workers and those of proprietors, although the latter prevail—and this not only affects the possibility of personal development for the individual but also the overall efficacy of the economy. Many cooperative endeavors are meant to foster values related to commitment, duty, and the fulfillment of one's professional potential, as advocated by Tawney.

We know a few successful modern institutions that are stable, socially flexible, and sensitive to the diverse motivations of their members. Universities, for example. Because of their many miseries, we rarely appreciate what interesting organizations they are. They are fairly autonomous and have unique characteristics, but they are not experimental or charitable institutions; in fact, they play a central role in complex societies. There is corruption, egoism, and a surprising number of quarrels within university organizations. There is also competition and cooperation, altruism and engagement, fraud and loyalty. Universities can be extraordinarily elitist or relatively egalitarian. They are not exactly bureaucratic state organizations, though many of the most important ones around the world are public. Some of them are private ventures and are even partially guided by monetarist criteria, though it is hard to imagine that they could perform their function if they were conventional businesses driven only by the profit motive.

In 1926, John Maynard Keynes argued that this kind of corporate organization was more present in modern societies than market ideologues would acknowledge. He was so convinced of their importance that he believed capitalist endeavors would end up looking like universities:

> The trend of joint stock institutions, when they have reached a certain age and size, [is] to approximate to the status of public corporations rather than that of individualistic private enterprise. One of the most interesting and unnoticed developments of recent decades has been the tendency of big enterprise to socialise itself. A point arrives in the growth of a big institution—particularly a big railway or big public utility enterprise, but also a big bank or a big insurance company—at which the owners of the capital, *i.e.* its shareholders, are almost entirely dissociated from the management, with the result that the direct personal interest of the latter in the making of great profit becomes quite secondary. When this stage is reached, the general stability and reputation of the institution are more considered by the management than the maximum of profit for the shareholders.[24]

It is hard to overstate just how mistaken Keynes was in his historical prognosis. On the other hand, his argument is a useful description of an institutional setting in which postcapitalist mechanisms and apparatuses could thrive. The proposal is not particularly attractive for those of us who hope that enlightenment, democracy, and technological progress might bring us more than a prêt-à-porter version of the Ivy League. The most interesting thing is how incendiary this proposal is, in spite of its restraint. Why does this relatively moderate scenario of institutional creativity seem so implausible to us today? Despite the devastating crisis of representation in Western political systems—according to polls, the Spanish see the political class as one of the five biggest social problems—any alternative, even one as restrained as Keynes's, is interpreted as political millenniarism.

In the mid–nineteenth century, Europe was haunted by the specter of communism. This is often taken as a bit of propagandistic bombast from Marx and Engels, but in fact it is a bit of poetic license taken in faithfully describing a political reality. Governments across Europe were bracing themselves for widespread revolt; the masses of workers were literally considered a dangerous class. This was a situation that, with subtle variations, lasted until World War II. In contrast, the specter most governments fear today is that of the toxic effects of their own economic and social policies, not an organized transition to a free and equal society. The political asthenia of the richest, most educated, and most informed societies in history is genuinely shocking. A similar lack of fortitude in the worlds of science, society, culture, art, or even sports would be unthinkable. Athletes do not just stop running because the records are too hard to beat, just as scientists did not shut down their labs after Max Planck came up with quantum theory.

Cyberfetishism and sociophobia are the final stages of a profound degeneration in the way we understand the sociality that shapes our notion of politics. We believe that we can satisfy our natural need to depend on other people, not only to live but also in the construction of our identities, through granular and limited relationships. We are far more dependent on others than, say, a group of hunter-gatherers, but we like to imagine ourselves as autonomous beings that cherry-pick from the forms of sociality on offer. The origins of this mutation, of course, predate digital networks. In fact, Internet-centrism has probably taken hold so quickly because it taps into a social dynamic that was already in place. The cornerstone of postpolitics is consumerism, the intertwining of our understanding of reality and generalized commoditization.

Consumerism is not the desire to acquire or flaunt objects but is rather a way of being in the world. We are consumers because we are only able to understand ourselves through some aspect of buying and selling. Our detailed understanding of social life is a byproduct of the market's infiltration of our bodies and minds. Consumerism is a way of internalizing inequality, in both the sense that we adopt it as part of our subjectivity and that—as we do so—we obscure it. In our fanatical submission to window displays, we intensify the importance of our personal choices and blur the line between these and class inequality.

Our definition of a home, for example, has changed radically. We are, in fact, nomadic societies, and our families are ridiculously small, but we dedicate far more resources to finding a place to live than any traditional sedentary society with far greater family units. We look for homes, but what we find are usurious mortgages, exploitation in the workplace and forced job mobility, and grotesque interior design. Despite all this, we are able to imagine that we are making long-term investments, building careers, and aesthetically transforming our living spaces. Our lives are bland copies of those of the elite, and we look down on those who are not on our level.

Even when we are not using our time to sell our labor or purchase goods and services, we dedicate ourselves to consumerist activities. Now that viewers have freed themselves from the tyranny of commercial television, thanks to the Internet, they spend their time watching commercial television online on an industrial scale. They even work for free, altruistically translating and subtitling series in order to do so. Freedom of choice has not helped us develop and appreciate new aesthetic forms but rather to mass consume what the market was already offering us—while seeing it, nonetheless, as an independent undertaking.

The conventional political sphere is defined by commerce, both in a descriptive sense—Hotelling's commercially inspired law elegantly sums up the monotony of political platforms—and in a regulatory one—the d'Hondt method applies the law of supply and demand to voter choice. Processes of emancipation have also been affected by consumerism. For example, Western girls come into virtually no contact with toys or objects that are not blatantly gendered. Princesses and fairies are a plague that have infected bibs, spoons, cradles, puzzles, books, blankets, pacifiers, potties, tricycles, and so on. Everything, absolutely everything, from the first day on is meant for a boy or a girl. As strange as this may sound, it wasn't always this way; the material world children inhabit has been transformed in recent decades. Some feminists view this as a step backward for egalitarian dynamics, a kind of neosexist counter-attack. This is a paradoxical argument, though, as it is hard to deny that gender equality, while still far from being complete, is greater than it has been at any other moment in modern history and, above all, that there is growing awareness among men that the shift toward equality is legitimate, positive, and irreversible.

More than anything, neosexism is a byproduct of consumerism. It is the result of intensively applying strategies of product differentiation to a basic anthropological reality like gender. The avalanche of gendered trinkets made for children is about turning them into compulsive consumers from the moment they are born (and even before, in fact).

The power of consumerism is fascinating. It is an incredibly ecumenical way of seeing things. Liquid subjectivities and identity-based communities call a truce every day to go to the mall to pick up smartphones and Adidas pants. From the Saudi Arabians who buy expensive Dior dresses to wear under their abayas, to the narcos who drive their SUVs around the slums where they live dressed like rappers, and the wealthy geeks who buy hybrid cars and rustic furniture; from the young men in Cairo who, after being inundated with pornography, harass women in staggering numbers during the festival to mark the end of Ramadan, to the high-altitude tourists who risk their lives to be shepherded up to the peak of Everest, and the urban cyclists who spend a fortune on minimalist bicycles without brakes or gears; from the children who wear soccer merchandise like uniforms, to the passengers who take luxury cruises on transatlantic ships the size of a skyscraper . . . The only thing that unites us is our ritualistic loyalty to buying and selling. No religion in the history of the world has achieved such universality.

In a way, the destructiveness of consumerism is surprising. Socialism sought to improve the material conditions of a great number of people who lived in abject poverty. Many thought that Fordism and the welfare state were simply capitalist versions of the same goal, an indirect attempt to spread material prosperity through mass consumption—the principal and essential limitation of which is the apparatus of capital accumulation itself.

This stance is overly generous toward the market. Consumerism rephrases the question that socialism was trying to answer as gibberish. Our lack of institutional apparatuses like central planning that could replace the market is something that we have to take seriously. We have capitulated to political immanence, to unending debates over the norms that govern the public sphere, including our own survival. Consumerism pulls the rug out from under us. It prevents us from structuring our desire for goods and services in a way consistent with the norms we believe should govern our communities.

To his great merit, the philosopher Walter Benjamin understood that mass consumption transforms not only what the market offers but also how we see the world. Benjamin, however, believed that this change could be managed to positive ends. Socialism would make more reasonable, more conscious use of steam power than capitalism did, converting it into a source of prosperity and equality. The same would go for the wares on display in shop windows and department stores. Benjamin believed that these things contained the seeds of liberation. Consumerism was the cultural complement of the material and political processes that socialism needed to transform and incorporate into its project.

This is not so outrageous. After all, some of the most innovative and successful companies of the past decade have developed models that are essentially monopolistic and centralized. IKEA, Zara, and H&M specialize in offering products that look much like those of designer brands and are of similar quality, but at a significantly lower price. It is not hard to imagine these chains as a kind of advancement or socialist version of mass consumption, albeit an exploitative and alienating one.

Benjamin was interested in consumerist subjectivity because he saw it as a way of accessing a richer aesthetic and political sensibility than that of the nineteenth-century bourgeoisie. He believed that faith in historical progress was one of the principal causes of political subjugation. The idea of progress essentially implies that history, in general, makes coherent sense, that is, that there are events that are inherently important or insignificant. The illusion that things work out the way they should, that the present is the inevitable result of the past, keeps us from considering the unrealized possibilities concealed by our reality. By breaking with the fantasy of progress, we gain access to a repository of alternatives aligned with our era, like radical political transformation.

Benjamin believed that the residents of the big cities where mass consumption was taking shape lived historically innovative experiences, whether they knew it or not. It was clear that they had left behind them the cyclical time of traditional societies, the rhythm of harvests and seasons. But there was also something wild and uncontrollable in the metropolis, something hard to reduce to an orderly narrative about civilization's advance. Consumers were in a historically privileged position—the dominant ideology spoke to them of progress—but their daily experience made them feel the discontinuous nature of reality, the universe of possibilities buried by the facts of their present moment. In a big city in permanent social, cultural, and material transformation, the idea of living a complete and definitive reality seemed almost absurd: "The destructive character sees nothing permanent. But for this very reason he sees ways everywhere. Where others encounter walls or mountains, there, too, he sees a way. . . . What exists he reduces to rubble—not for the sake of rubble, but for that of the way leading through it."[25]

This optimistic vision of consumer society runs up against a strange limit when the fragmentation of experience emblematic of capitalist postmodernity comes to dominate discourse. We live in a semiotic jungle that rewards fragmentation and punishes continuous, coherent narratives. The idealization of the Internet and of digital communities is the most patent ideological manifestation of this. Advertising has become a polymorphous strategy that engenders its own complex and sometimes ironic games. The most effective marketing strategies are those that eliminate the distance between the sender and the receiver of the message not by means of traditional empathy but rather through some simulacrum of cooperative creation.

As an environment, liquid modernity is extremely hostile to those who seek to construct a solid identity, a consistent subjectivity based on a teleological narrative. Those who triumph within turbocapitalism are profoundly adaptive: They have different selves—varied domestic, ideological, and professional personalities. Those who are vanquished by it, as well. Migrant workers no longer go to another country with the idea of starting a new, more prosperous life; instead, they spread their labor around, jumping from country to country as they follow the capricious flow of capital. Even therapists exhort us to accept this extreme liquidity. Whoever latches on to a political, romantic, or moral identity is no longer just a loser or resentful; they are now also pathologically inadaptive. The dominant social and political metaphors of our time have to do with reticularity and fragmentation: networked society, distributed systems, modular minds.

Walter Benjamin underestimated the nihilistic charge behind consumerism that cyberfetishism unequivocally brings to light. Consumption leaves behind not ruins but trash. The destructive character of our day needs to find his way through the dump. Benjamin was wrong because he never got to know the millenniarist models of postmodern consumerism, the way it kills all hope of reconciliation with deep-seated anthropological forces. Consumerism is to the sensibility of modern society what casino capitalism is to its economy. But this failure is interesting because it reveals an important limitation of political transformation. Emancipation might be compatible with some forms of market or bureaucracy but not with consumerist inequality or anything derived from it, like cyberfetishism and sociophobia.

Strangely enough, the first thinkers to condemn the destructive potential of consumerism were written off as elitists, even by the Left. Christopher Lasch and Pier Paolo Pasolini saw clearly that consumerism implied an acceleration of the closing off of historical possibilities that Benjamin sought to correct:

> The right of the poor to a better existence had a counterpart which has ended by degrading them. The future is imminent and apocalyptic. Sons are snatched away from similarity to their fathers and projected towards a tomorrow which, while preserving the problems and miseries of today, cannot but be qualitatively

different. . . . The break with the past and the lack of rapport (even if ideal and poetic) with the future are radical.[26]

There is simply no such thing as a shared life when life is viewed through a shop window. Or in the digital age. In fact, to the extent that cyberfetishism is based on the appearance of abundance, it has exacerbated the problem by decoupling consumerism and trade. On the Internet, consumerism has finally revealed itself as the devastating historical force that it is. Today, we can experience the alienation of the consumer, even without the mediation of money.

◎ ◎ ◎

There is a subtle ideological connection between the false promises of the social sciences, the institutional rigidity that paralyzes political change, and the dissolution of community relations. Its foundation is the illusion that social reality and our ideas about it are clearly defined, that they are finite, analyzable facts with a precise conceptual architecture. As though social processes had a molecular structure that could be configured in different ways, preferably by means of a spontaneous impulse toward self-organization or, failing this, through centralized planning.

This sociological chimera has had profound effects on our perception of the public sphere. Specifically, our understanding of social inequality is completely contaminated by the failure of the social sciences. For decades, there have been attempts to define the concept of social class with the greatest precision possible, using lavish theoretical trappings. It is the story of a persistent failure, as there is always some group that refuses to fall between these rigid lines. They might be salaried workers with high purchasing power, or entrepreneurs without much cultural awareness, or unsalaried housewives, or workers who have a say in how their workplace is run, and so on. As such, in a kind of academic version of the Procrustean bed, sociologists, economists, and political theorists have decided that class inequality is no longer that important in a globalized world made up of social networks in constant flux. And we've believed them. We like to imagine ourselves as sophisticated actors in a well-distributed system of information and communications, not as precarious, submissive laborers obsessed with brand-name trinkets. In reality, a concept of class based on broad, inexact criteria—income, control over one's own working conditions, social prestige—turns out to make more sense than ever and is indispensable in today's world to understanding who wins and who loses, and to what extent. This line of argumentation, however, is not easily accommodated by sophisticated social theories desperately seeking conceptual precision, even at the cost of empirical content.

In general, the moral and social universe suffers from profound ontological relativity, to borrow a phrase made popular by the philosopher W. V. O. Quine. It is populated by hazy, imprecise realities without clearly defined limits, about which we have confused, inexact ideas. We are condemned to intervene in these realities using a stream of prudent conceptual apparatuses and to understand them through contingent explanatory mechanisms. Why, then, do we still bow to economists and psychologists who speak to us of imaginary entities? To politicians whose words we no longer register, much less believe? Why do we refuse to acknowledge the damage to our own lives and idealize the dual crutches of psychopharmaceuticals and technology? The answer, at least in part, is consumerism. It is a simple ideology based on the mechanics of desire, but it is also far-reaching and highly effective.

We all understand that freedom, equality, and personal development are goals run through and through with indeterminacy. They are dispositional realities, rather than facts. It is like stating that someone knows Italian: We are not saying that something is happening to that person at a given moment but rather that they are able to do certain things if the occasion calls for it. We must be constantly developing our political values—because we change them and because they change us as we try to pin them down. Often, this process can only happen as a community. True equality, for example, is not a starting point but rather a goal. Sentimental egalitarian declarations like "we are all the same" are superficial, even counterproductive. We are not the same. In fact, we are quite different. That kind of sameness, or equality, is a political invention, a byproduct of the construction of citizenship and democracy that must be systematically cultivated.

Consumerism, on the other hand, offers us a comforting concreteness. It is the kind of activity in which the goals are clearly established and there is no use debating them. It consists simply of choosing the means I think will best satisfy my desires. Adidas or Nike? Windows or Mac? There is nothing inherently wrong with this. Our daily lives would be impossible if we subjected every little choice to critical examination. The problem arises when this kind of activity takes on a significant symbolic charge and becomes a privileged source of meaning. When it becomes the space in which our individual identities are forged.

Consumerism is a meager but immediate form of satisfaction that, because it is carried out through a quantitative operation, seems to be clearly defined. As a result, we want our understanding of all of social reality to have the same degree of analytical precision as this one overdetermined situation. In the market, our interactions are simple, finite, and easily conceptualized. Why not explain the rest of our lives with the same precision and simplicity? We vote for what we want. We like to drive. Our socialization can be measured in bits. By integrating our bodies into the mechanics of the market, we legitimize the explanatory false promises of the social sciences.

In cyberfetishism, consumerism achieves consciousness: It is no longer simply the symbolic background noise of capitalism but rather a social and cultural project. Cyberfetishism is the political coming of age of consumerism. In the minds of cyberutopians we are, finally, no longer alone in the city, condemned only to run into one another from time to time in line at the supermarket. We believe that we have overcome the malaise of material prosperity, the problems of Fordist individualism, and the forms of alienation associated with these. We see ourselves as clusters of intermittent but intense choices adrift in the reticular circuitry of postmodern globalization. We are fragments of personal identity that collide in digital and analog social networks.

The price we pay is the destruction of all projects that require a real commitment. According to cyberfetishism, we are no more than our current social, gastronomical, musical, sexual, cinematic, or even political appetites. The modern age has experienced this dissolution of subjectivity into volitional content as a form of nihilism that, in the long term at least, has generated suffering and malaise, as is the case of those middle-class housewives in *Mad Men* who spend their days numbed by sedatives. The digital sphere offers us a technological crutch that stands in for stability in our irregular choice making. The Internet creates an illusion of intersubjectivity that nonetheless does not mean engaging with norms, people, or values.

This is why ours is the age of the scientific failure of the social sciences as well as the moment of its greatest cultural success. No one knows who the big names in sociology, economics, or education were, anymore. Skinner? Galbraith? Dahrendorf? Doesn't ring a bell. Yet we behave as though the dean of some program in social sciences were running our lives. If a committee of rational-choice theorists, psychoanalysts, and educators had been asked to reach some kind of minimum consensus and start a project for social relations, they would have invented Facebook. The same is true of public policy. No one in their right mind thinks that conventional governments are able to develop a coherent political strategy capable of anything more than obsessively destroying the ruins of Keynesianism. Nonetheless, rarely in recent history has there been such vertigo in the face of political innovation, such violence against any challenge to the disaster in which we are entrenched, such sociophobia.

Projects of political emancipation are exactly the opposite: They are the institutional realization of substantive ethical endeavors. These ventures are not hollow ones; these are not metaprojects. As Tawney explained, they are more firmly grounded in duties and obligations than in rights. For socialists, the question is constructing those duties and obligations that will commit us to overcoming material inequality, state paternalism, and alienation.

The complexity of our political reality demands that we break with the legacy of social science's false promises, that sophisticated and soporific form of intellectual consumerism. Practical wisdom in public matters is built on long-term deliberative processes; it is not the patrimony of tribunes, experts, or nobles. The greatest challenge that radical democracy faces in postmodern times is not confusing itself with the choices made by consumers in the market or by users online. Political reflection has nothing to do with the consistent aggregation of preferences through some technological device, whether in commerce or on a social network. Collective deliberation is a process by which shared objectives are constructed and not a mechanism to make a set of given options—which may be partially or totally opposing—more compatible with one another.

This is why the ethics of caregiving is so richly political. Not because politics resembles family relations—in a sense, they are exactly the opposite—but rather because in the sphere of caregiving it becomes clear just how much the norms we embrace turn us into people who can aspire to be different from how we are, and sometimes we can only achieve that as part of a group. Democracy cannot be broken down into bundles of isolated decisions because it involves the commitments that define us as individuals with some degree of consistency, with a past, and with some vague expectation of a future. And this anthropological reality is incompatible with cyberfetishism and sociophobia.

CODA

1989

I ONCE SAW a television documentary about the social reintegration of Latin American paramilitary soldiers who had laid down their arms. The film showed something like acts of reconciliation in which the criminals offered explanations to, and asked forgiveness of, the families of their victims. There was one man who said, by way of an excuse, that he had killed many people but had never cut anyone's head off with a chainsaw, like some of the others had. I remember thinking that if the best thing you can say about yourself is that you've never beheaded anyone, you really need to simplify your moral life.

Many postcapitalist proposals inspire a similar reaction in me. The best thing they seem able to say about themselves is that they are not impossible—and even that is asserted without much enthusiasm. Cyberfetishism, sociophobia, and other byproducts of consumer culture have left their mark on our perception of our political reality. Social transformation, understood as a realistic project and not just an activity that provides bit of comfort for the beautiful people, turns out to be overwhelmingly complex—but no postcapitalist project can call itself such if it is not prepared to deal honestly with this complexity.

My political experience is quite limited, to be honest. When I was fifteen, just a few months after the Berlin Wall came down, I joined a non-Soviet communist youth group. It did not last long, but I consider it an important moment in my political education. There are communists today who still announce the imminent *sorpasso* of the global proletariat on a daily basis. This is a fallacy of composition: They add up many small strikes and microrevolts from around the world and get a revolution on a global scale. I saw none of this in my organization. On the contrary. Discussions of the imminent political future oscillated between abject pessimism and irony. The sense that we were part of a dying project hung in the air.

Unlike the antimilitarist movement, in which I was also an active participant, I have only a vague memory of my brief communist activism. This is strange because the experience involved lots of issues and interminable meetings. If I am not mistaken, a good deal of time was dedicated to coming up with ideas to avoid the dissolution of the party. This drove me crazy, as it would have done to any hotheaded teenager. I believed that political action would automatically have an effect on ties within the organization. If we just did what we had to do—whatever that was, almost certainly nothing reasonable—things would get better.

In retrospect, I can imagine that I witnessed a very interesting process, though I was of course unable to appreciate it at the time. Our attempt to survive as an organization, which was ultimately unsuccessful, made a lot of sense to the older members of the group. At stake was not only their identity, in which activism played an important role, but also personal relationships forged over decades of intense political action. Those bonds would not survive, or they would be greatly weakened, without the support of a collective actor that gave them meaning. The strange thing is that this awareness of the importance of personal connections had barely any place in the group's activism. Only when the institutional crisis proved irreversible did it come to light that there was an important component to activism that had to do with a bond of fraternity that is not formed under other circumstances.

A while later, I participated in various projects related to open culture. In that context, community and cooperation were fetishized terms; nonetheless, their effect was quite limited. The kind of personal commitment that was a given among traditional left-wing activists, who handed over part of their salary every month to the organization, was barely present here. The idea of reporting on the outcomes of one's activity to someone else was met with a mix of incomprehension and mistrust. You did what you could, with the best attitude you could muster, more or less when you felt like it. The practical results of this were almost always calamitous, but what surprised me most was that there was significantly more hostility that you might have expected to find in that empire of free will. And this is coming from someone who was more than used to impassioned discussions on how to end capitalist relations of production. There seemed to be an inverse relation between political engagement and personal aggression. This is very clear in the case of digital cooperative movements, in which highly technical and often trivial questions of licensing, protocols, and formats often lead to major dialectical battles. It seemed to me that this cybernetic irritability was the symptom of the political fragility of social technologies that, in turn, epitomized postmodern social connections.

The 15-M movement had a major effect on me. It was as though postpolitics was falling apart before my eyes, not to return to modernity but rather to reconfigure what had been left to it. A call to arms that at first sounded more like a flash mob than anything else evolved in the space of barely a week to take on a significant role in the anticapitalist project. And it had an incredible reach. The Saturday following May 15, 2011, I took the subway to Plaza del Sol, in the center of Madrid, late in the evening. The train was full of teenagers heading to the bars downtown, as they did every weekend. It was an incredible experience: Every one of them seemed to be talking about politics . . . as though in just a few weeks the wall of cynicism that sentences us to live damaged lives had crumbled. For the first time, political arguments—which were, at times, naïve, skewed, or populist—occupied an explosive symbolic space that for recent decades had been dominated by ringtones, ridiculous and expensive clothing, soccer, amateur porn, and cat videos.

This may be why so many people misunderstood the relationship between 15-M and the Internet. Many have thought that information technology was a catalyst in these political processes.

I believe exactly the opposite to be true. The 15-M movement followed such a winding road because it had to overcome the vicious impasse created by consumerist cyberfetishism. The Internet has become an important weapon—not for getting people out into the streets but for use when they are already out there. We have been forced to abandon the idea that intervening in the public sphere is a matter of writing reactionary messages in some online forum. Opening the door for someone at the supermarket when we are in a good mood is no longer our most intense form of social interaction. If the Internet is playing an important role in this movement, it is because we have rediscovered the power of face-to-face encounters and commitment, and we have come to understand that we live in a context in which these things are extraordinarily hard to come by.

I believe that diehard left-wingers like Christopher Lasch were correct in asserting that it was misguided to accept capitalism's destruction of traditional social bonds as good news; that's a bit like being contracted to demolish a building and arriving to find that the job has already been done. Rather, socialism is like the boat described by Otto Neurath, who said we are all sailors who need to rebuild their ship while still at sea. Unlike the communitarians, however, I do not believe that the question of community is relevant when it comes to proposing solutions. Deep social relations prosper spontaneously when the conditions neutering them are removed. This might be a confusing, difficult, or painful process, but it is not exactly a political problem. For example, extended families are beginning to reappear as a result of postmodern fragmentation. Not because of a conservative return to traditionalism but rather the opposite: second marriages, sexual diversity, job mobility, and the need for support in the wake of the financial crisis are causing a rise in the number of unconventional polynuclear families.

Institutional design, on the other hand, is the opposite of spontaneous: It requires deep and sustained pragmatic deliberation that cannot be resolved using formalist theoretical devices. In cyberfetishism these things are turned on their head—true sociality seems to require an extensive technological apparatus, while the shape of institutions seems left to chance.

Sometimes when I hear about a new oppositional initiative, I ask myself whether I would like to see the individuals involved in positions of responsibility in a noncapitalist society. The answer, most of the time, is that I wouldn't trust them as either administrators or as part of my community of neighbors. On other occasions—very few, to be honest—the answer is yes, absolutely. This may seem strange, but the Left has not been an environment especially conducive to these questions, which are easy to ask but much more complicated to answer—and which are, in my opinion, the very essence of a nonrhetorical political practice. In the digital sphere, and in the context of hyperconsumerism, these questions are simply meaningless.

Beginning on May 16, 2011, a bunch of friends started writing and calling to say that 15-M had refuted or confirmed various political and sociological theories, which for the most part already contradicted one another. Followers of Negri asked me irately if the idea of the multitude as the new revolutionary subject still seemed funny to me. The anarchists scolded me for my institutionalist skepticism of direct democracy. The Leninists underscored the persistence of the class struggle I had questioned before. Last of all, the postmodernists emphasized how 15-M deconstructed modern political experience and did away with the great emancipatory metanarratives.

None of them managed to convince me, of course. Instead, I realized how deeply I felt that, to say it in something of a trite way, we were participating in the process of becoming what we already were. One of the questions behind this book is, precisely, how some of the problems of modernity related to political emancipation and social relations persist in our age of digital screens and megaslums, of social networks and *pisos patera*—those overcrowded, illegal apartments that are often the only housing option for migrant workers. I believe that the answers traditional oppositional politics offers to these problems are no longer of use and at the same time are indispensable.

Revolutionary traditions have squandered part of their experience because they have misunderstood themselves as the fever dreams of some drunken sociologist. Perhaps now that the rest of the world sees them as all washed up, it might be a good moment to correct this error and think about postcapitalism as a feasible and agreeable project within our reach. I believe that tackling the immense complexity of such an aspiration is essential, though the ecosystem of cyberfetishistic consumerism puts enormous pressure on us in the opposite direction: Typing 140 characters while dressed like a clown in the latest trends is the new final frontier of banality. Changing the world might be hard, but it is not necessarily complicated. A viable form of postcapitalism, however, is infinitely complex. Just as complex as the day-to-day reality of the relationships within our communities that we do not, and will never, fully understand.

NOTES

GROUND ZERO: SOCIOPHOBIA

1. Mike Davis, *Late Victorian Holocausts: El Niño Famines and the Making of the Third World* (New York: Verso, 2002).

2. Eric Hobsbawm, *The Age of Empire, 1875–1914* (New York: Pantheon, 1987), 59.

3. Karl Marx, *New-York Daily Tribune*, June 25, 1853.

4. Davis, *Late Victorian Holocausts*, 9.

5. Mike Davis, *Planet of Slums* (New York: Verso, 2006), 23–26.

6. Rose George, *The Big Necessity* (New York: Metropolitan, 2008), 2.

7. Sven Lindqvist, *A History of Bombing* (New York: New Press, 2001), 86.

8. Kurt Vonnegut, *Player Piano* (New York: Scribner, 1952; repr., New York: Delacorte, 1971), 19.

9. Milton Friedman, *Capitalism and Freedom* (Chicago: University of Chicago Press, 1962; repr., Chicago: University of Chicago Press, 1982), 22–24.

10. Bill Bryson, *At Home: A Short History of Private Life* (New York: Doubleday, 2010), 217.

11. Paul Collier, *The Bottom Billion* (New York: Oxford University Press, 2007), 28.

12. Göran Therborn, *The World: A Beginner's Guide* (Cambridge: Polity, 2011), 178.

13. Steven Lukes, *The Curious Enlightenment of Professor Caritat* (New York: Verso, 1995), 191.

1. DIGITAL UTOPIA

1. Le Corbusier, *Aircraft* (London: Trefoil, 1935), 85.
2. Lewis Mumford, *Technics and Civilization* (Chicago: University of Chicago Press, 2010), 215.
3. David Noble, *America by Design* (New York: Knopf, 1977), 16.
4. Evgeny Morozov, *The Net Delusion* (New York: Public Affairs, 2011), 71.
5. Igor Sádaba, *Propiedad intelectual. ¿Bienes públicos o mercancías privadas?* (Madrid: Catarata, 2008), 57ff.
6. Erik S. Reinert, *How Rich Countries Get Rich . . . and Why Poor Countries Stay Poor* (New York: Carroll & Graf, 2007), 186–188.
7. David Ariestegui, "Un ministerio de cultura en la sombra," in *CT o la Cultura de la Transición* (Madríd: Debolsillo, 2012). See also Ariastegui, "Capitalismo de casino y derechos de autor," http://info.nodo50.org/Capi talismo-de-casino-y-derechos.html.
8. Yeyo Balbás, "Negocio de reseñas 2.0," *Cultura libre*, http://www.cultura libre.org/negocio-de-resenas-2-0/.
9. Cory Doctorow, "De cómo los derechos de autor deberían cambiar para ajustarse a la tecnología," *Minerva* 9 (2008), http://www.revistaminerva .com/articulo.php?id=278.
10. Jaron Lanier, *You Are Not a Gadget* (New York: Vintage, 2011), 182.
11. "The GNU Project," GNU Operating System, last modified December 1, 2015, http://www.gnu.org/gnu/thegnuproject.en.html.
12. Toni Domènech recounts this anecdote in "Después de la utopía. Coloquio entre Antoni Domenech y Daniel Raventós," *Minerva* 15 (2010): 59. http:// www.circulobellasartes.com/revistaminerva/index.php?id=19.
13. Morozov, *The Net Delusion*, 63ff.
14. Montesquieu, *The Spirit of Laws* (Edinburgh, 1768), 2.20.1, 2.21.20.
15. Lanier, *You Are Not a Gadget*, 34.
16. Dan Ariely, *Predictably Irrational: The Hidden Forces That Shape Our Decisions* (New York: HarperCollins, 2008), 71.
17. Jon Elster, *Closing the Books: Transitional Justice in Historical Perspective* (New York: Cambridge University Press, 2004), 109–110.
18. Kwame Anthony Appiah, *Experiments in Ethics* (Cambridge, Mass.: Harvard University Press, 2008), 132–133.
19. Elinor Ostrom, *Governing the Commons: The Evolution of Institutions for Collective Action* (New York: Cambridge University Press, 1990), 51.
20. Ibid., 60.
21. Ibid., 183–184.

2. AFTER CAPITALISM

1. Slavoj Žižek, *Welcome to the Desert of the Real!* (New York: Verso, 2002), 5.
2. John Berger, *A Painter of Our Time* (New York: Vintage, 1996), 198.
3. Gerald Cohen, *If You're an Egalitarian, How Come You're So Rich?* (Cambridge, Mass.: Harvard University Press, 2001), 101.
4. Walter Benjamin, "Bert Brecht," in *Selected Writings*, vol. 2: *1927–1934* (Cambridge, Mass.: Harvard University Press, 1999), 369.
5. James E. Miller, *The Passion of Michel Foucault* (Cambridge, Mass.: Harvard University Press, 2000), 203. [Translator's note: The bracketed sentence does not appear in the English edition, but it is present in the Spanish.]
6. Erik Olin Wright, with Andrew Levine and Elliott Sober, *Reconstructing Marxism* (New York: Verso, 1992), 148–149.
7. Bertolt Brecht, *Poems 1913–1956*, ed. John Willet and Ralph Manheim (New York: Methuen, 1976), 318.
8. John Elster, *An Introduction to Karl Marx* (New York: Cambridge University Press, 1986), 45.
9. Stephen Mulhall and Adam Swift, *Liberals and Communitarians* (Cambridge, Mass.: Blackwell, 1992), 15.
10. Dan Ariely, *Predictably Irrational: The Hidden Forces That Shape Our Decisions* (New York: HarperCollins, 2008), 76–77.
11. Alasdair MacIntyre, *Dependent Rational Animals* (Chicago: Open Court, 1999).
12. Lauren Slater, *Opening Skinner's Box: Great Psychological Experiments of the Twentieth Century* (New York: Norton, 2004), 60.
13. Judith Rich Harris, *The Nurture Assumption* (New York: Free Press, 1998).
14. Kate Pickett and Richard Wilkinson, *The Spirit Level: Why Greater Equality Makes Societies Stronger* (New York: Bloomsbury, 2009), 54ff.
15. Evgeny Morozov, *The Net Delusion* (New York: Public Affairs, 2011), 184ff.
16. Manuel Pérez, in Hans Magnus Enzensberger, *Der kurze Sommer der Anarchie; Buenaventura Durrutis Leben und Tod* (Frankfurt am Main: Suhrkamp Verlag, 1972), 95–96.
17. Betty Friedan, *It Changed My Life: Writings on the Women's Movement* (Cambridge, Mass.: Harvard University Press, 1998), 397.
18. Alasdair MacIntyre, "Is a Science of Comparative Politics Possible?" in *Against the Self-Image of the Age* (Notre Dame, Ind.: University of Notre Dame Press, 1978), 260.
19. Naomi Klein, *The Shock Doctrine* (New York: Picador, 2008), 67–68.

20. Alec Nove, *The Economics of Feasible Socialism* (New York: Routledge, 1983), 74.
21. Francis Spufford, *Red Plenty* (Minneapolis, Minn.: Graywolf, 2010).
22. Bernard Gerrien and Sophie Jallais, *Microeconomía: una presentación crítica* (Madrid: Maia, 2008), 32–34.
23. Jon Elster, *Nuts and Bolts for the Social Sciences* (New York: Cambridge University Press, 1989), 9.
24. John Maynard Keynes, "The End of Laissez Faire," in *Essays in Persuasion* (New York: Norton, 1963), 314.
25. Walter Benjamin, "The Destructive Character," in *Selected Writings*, vol. 2: *1927–1934* (Cambridge, Mass.: Harvard University Press, 2005), 542.
26. Pier Paolo Pasolini, *Lutheran Letters*, trans. Stuart Hood (New York: Carcanet, 1987), 36.

INDEX

INSURRECTIONS: CRITICAL STUDIES
IN RELIGION, POLITICS, AND CULTURE

SLAVOJ ŽIŽEK, CLAYTON CROCKETT,
CRESTON DAVIS, JEFFREY W. ROBBINS, EDITORS

GPSR Authorized Representative: Easy Access System Europe, Mustamäe tee
50, 10621 Tallinn, Estonia, gpsr.requests@easproject.com

www.ingramcontent.com/pod-product-compliance
Lightning Source LLC
Chambersburg PA
CBHW032137020426
42334CB00016B/1200